Turfgrass
WATER CONSERVATION

Technical Editors

Victor A. Gibeault
Extension Environmental Horticulturist
University of California, Riverside
and
Stephen T. Cockerham
Superintendent of Agricultural Operations
University of California, Riverside

Cooperative Extension
UNIVERSITY OF CALIFORNIA
Division of Agriculture and Natural Resources

Contents

Introduction — 4

1 The size, scope, and importance of the turfgrass industry

Factors influencing turfgrass increase / functions of turf / turf in a hypothetical city / the national turfgrass industry / conclusions — 7

Stephen T. Cockerham, Agricultural Operations, University of California, Riverside, CA 92521

Victor A. Gibeault, Environmental Horticulture, Cooperative Extension, University of California, Riverside, CA 92521

2 Water: Whose is it and who gets it?

How much water is there? / who owns the water? / who manages water? / who uses water? / what are the priorities? / what are the options? — 13

Joseph P. Rossillon, Freshwater Foundation, 2500 Shadywood Road, Navarre, MN 55392

3 Water resources in the United States

Practicum — 19

Water uses / hydrologic cycle / water sources and supply / precipitation / runoff water / surface water / groundwater / overdraft and its consequences / depletion / alternatives / groundwater implications for turfgrass / water resources, regions, and subregions / summary — 21

James R. Watson, The Toro Company, 8111 Lyndale Avenue South, Minneapolis, MN 55420

4 Physiology of water use and water stress

Practicum — 37

Water potential / water movement in unstressed plants / water-use rate / water deficit and water stress / drought and drought resistance / efficiency of water use / general responses to water deficits / effects of water deficits on physiological processes / summary and conclusion — 39

Victor B. Youngner, formerly of the University of California, Riverside

5 An assessment of water use by turfgrasses

Practicum — 45

Plant/water relations / evapotranspiration / evapotranspiration measurement / resistances to evapotranspiration / controls of evapotranspiration / interspecies evapotranspiration characterizations / intraspecies evapotranspiration characterizations / plant improvement—breeding / research needs / acknowledgments — 47

James B. Beard, Soil and Crop Science Department, Texas A & M University, College Station, TX 77843

6 Turfgrass culture and water use

Practicum — 61

Cultural practices / chemicals / summary — 63

Robert C. Shearman, Horticulture Department, University of Nebraska, Lincoln, NE 68583

7 Influence of water quality on turfgrass

Practicum — 71

Common sources of water / water treatment / effluent water / water quality parameters / potable water / physical quality / solving problems associated with poor-quality water / leaching of salts / blending water / salt-tolerant grasses — 73

Jack D. Butler, Horticulture Department, Colorado State University, Fort Collins, CO 80521

Paul E. Rieke, Crop and Soil Sciences Department, Michigan State University, East Lansing, MI 48824

David D. Minner, Horticulture Department, 1–40 Agriculture Building, University of Missouri, Columbia, MO 65211

8 Soil/water relationships in turfgrass

Practicum — 85

The basics of soil / soil-plant-atmospheric system / soil characteristics affecting water management / irrigation programming techniques — 87

Robert N. Carrow, Department of Agronomy, University of Georgia, Experiment, GA 30212

9 Irrigation systems for water conservation

Practicum — 103

Budget considerations / efficiency of coverage / efficiencies of distribution / site considerations / design considerations / use of the irrigation system / avoiding problems — 105

Jewell L. Meyer, Cooperative Extension, University of California, Riverside, CA 92521

Bruce C. Camenga, Oasis Irrigation, 2001 Third Street, Suite E, Riverside, CA 92507

10 Influence of water on pest activity

Practicum — 113

Water's influence on turfgrass growth and microclimates / turfgrass weeds / fungal diseases / insects and related turfgrass pests / nematode diseases / pest control strategy with limited water resources — 115

Phillip F. Colbaugh, Texas A & M Research Center, 17360 Coit Road, Dallas, TX 75252

Clyde L. Elmore, Botany Extension, University of California, Davis, CA 95616

11 Site design for water conservation

Practicum — 131

Establishing design criteria / irrigation master planning / implementation / maintenance guidelines / water management programs / conclusion — 133

Cal O. Olson, Olson Associates, 19600 Fairchild, Irvine, CA 92713

Glossary — 150

Introduction

Turfgrass is an important component of our modern United States society because it has a direct effect on the way most people live. Many recreational facilities depend on a uniform, vigorous, well-maintained turf sward as the medium of play. Gardening is important to many people, and lawn maintenance is a constant source of challenge and pride for the home gardener. Turf also affects people's lives when used in ornamental settings to create the desired aesthetic appearance. And, of course, turfgrass directly influences our environment by reducing localized temperature, glare, noise, soil erosion, and chemical and particulate air pollution.

Water is necessary for turfgrass growth and development. Water is absorbed by turfgrass roots and transported upward through the plant where it is used for plant metabolism; it also functions as a transport medium for plant nutrients and foods. Lastly, most absorbed water is transpired from the turfgrass leaves to the atmosphere, a process which regulates the plant temperature. Without water all turfgrass plants would die.

Turfgrass receives water from either natural precipitation or irrigation. When natural precipitation is of insufficient amount and/or of inadequate frequency, turfgrass must be irrigated to provide the desired turf appearance and recuperative ability. Problems arise when there is an extended period of lack of precipitation and insufficient amount or availability of either groundwater or surface water to allow for turf irrigation.

The importance of water for turfgrass irrigation was most clearly realized during the drought periods of 1976–77 and 1977–78 in the western United States. Questions about turfgrass survival under water stress were asked; unfortunately, answers were not always obtainable because of insufficient research and because a comprehensive source of information was not available. Ideas for a turfgrass water conservation symposium and this resulting book were then formulated.

The symposium

The symposium, held on February 15 and 16, 1983, in San Antonio, Texas, was financially sponsored by the American Sod Producers Association (ASPA). ASPA was interested in this endeavor because of its need for turfgrass water-use data for its membership and also because of its desire to provide information to the total turfgrass community, including civic and political leaders.

The symposium was planned so that each presentation would become a chapter in this book. Using water conservation as the symposium theme, topics were selected for the individual chapters, and speakers/authors were invited according to their respective expertise.

The chapters were written and all of them sent to each author for review before the meeting. The symposium attendees consisted of the authors and a small number of guests, including ASPA board members, well-known turf authorities, and industry journalists. During the 2-day meeting each author presented a summary of his chapter, and because all authors had reviewed each chapter before the meeting, lively informal discussions took place. During the latter part of the symposium, comments and questions from the guests were taken and discussed.

After the symposium, the authors were given time to rewrite their chapters so that they could incorporate comments or suggestions on their presentations that would enhance their respective chapters. In addition to this rewrite, the authors were asked to provide a practicum for their chapters.

The practicum

The practicum is an important part of each chapter. Professionally, researchers must present their information in a technical manner, using data to substantiate their conclusions and recommendations. Often it is difficult for a less technically trained reader to derive and apply the information presented to a field situation.

To expand the value of this book to the reader, each author of a technical chapter has written a summary of his information specifically for practical use. Each summary is separate from the chapter and is designated as the "practicum."

The target audiences

Chapters 4 through 11 are aimed at the needs of landscape architects, designers, contractors, suppliers, and consultants. Sod growers and turf managers will find those chapters, plus chapters 2 and 3, to be of interest. Chapters 4, 5, 6, 7, and 8 can be helpful to researchers for background information. Educators and students in turf management will find there is something for them in each chapter.

Two of the interest groups, political and civic leaders, are not the usual target audiences for turf publications. Because of the impact of water shortages and conservation programs on society, political decisions must be made which affect water management, specifically turfgrass water management. In the past, unfortunately, political decision-makers have not had adequate information about the turfgrass industries available to them. Chapters 1, 2, 9, and 11 have been included to provide just that kind of information.

Acknowledgments

This book would not have been possible but for the board of trustees of the American Sod Producers Association. Much assistance was given by former ASPA executive director Bob Garey and current executive director Doug Fender. Shirley Potter of the ASPA staff handled the symposium arrangements in her usual efficient manner.

Kathy Weeks of the University of California, Riverside, was the person upon whom the miscellaneous jobs fell. Her dedication and performance were exceptional. Heidi Seney at the University of California Agriculture and Natural Resources Publications was key to editing and publication preparation. Marvin Ehrlich provided design and layout effectively. Forrest Cress of University of California, Riverside, offered specific editorial assistance.

Kathy Copley, editor of *Grounds Maintenance* magazine, was of tremendous assistance in editing and laying out the practica. Also, she attended the symposium as a guest and made several significant contributions. Along with ASPA board members, others guests who attended the symposium and played a vital part were Dr. John Addink, Steve Briggs, Dr. Henry Indyk, and Howard Kaerwer.

From the very early planning stages of the symposium through the writing of a chapter, Dr. Victor B. Youngner was active in this project. His untimely death occurred before the book was finished. Dr. Youngner's contribution is immensely appreciated.

Lastly, but most importantly, the authors were an excellent group to work with. Each of them took a personal interest in the success of this project, and we recognize that this book reflects the combined efforts of these very busy people.

Even though the authors are responsible for their respective chapters, the editors assume responsibility for the total content of this book.

Victor A. Gibeault
Stephen T. Cockerham
June 15, 1985
Riverside, California

1 The size, scope, and importance of the turfgrass industry

Stephen T. Cockerham and Victor A. Gibeault

The turfgrass industry in the United States is made up of individuals and groups involved in the design, installation, maintenance, sales, and service of turfgrass, and the educational and developmental aspects of this industry. Its expansion has largely occurred since World War II and can be attributed to the continued urbanization of the U.S. population and to increases in leisure time, discretionary income, demand for (and ability to pay for) amenities, and population.

STEPHEN T. COCKERHAM

Stephen T. Cockerham received his B.S. degree from Purdue University and his M.S. degree from New Mexico State University. Presently, he is the superintendent of agricultural operations, Citrus Research Center and Agricultural Experiment Station at the University of California, Riverside. Mr. Cockerham is past president of the American Sod Producers Association.

Factors influencing turfgrass increase

Urbanization. During the past 100 years, the U.S. has shifted from an agricultural society, with most residents living on farms, to an industrial society, with most residents living in urban, suburban, or small town locations. According to A. Toffler (21), we are now entering an informational society which will also have an urban, suburban, or small town population structure. Although people have been "freed" from the toil of an agricultural society, they are now removed from intimate contact with nature. Schery (19) notes this with the following:

> By observing the seasonal behavior of grass, one touches on the grand rhythm of natural events. With a grassy lawn at our dooryard, there is not only respite from the tension and press of the city, but instruction in biological cause and effect. Watching grass respond to soil and season may be, for city people, a last link to the solace and understanding our vanishing wilderness once gave.

Leisure time. Human life, as related to time, can be viewed in three categories: existence time, subsistence time, and leisure or discretionary time.

Existence time. Time spent to stay alive—eating, sleeping, and caring for one's health—is existence time. As a percentage of total human time, it has not changed much throughout human history.

VICTOR A. GIBEAULT

Victor A. Gibeault received his B.S. and M.S. degrees from the University of Rhode Island and his Ph.D. degree from Oregon State University. Since 1969 he has been an Extension environmental horticulturist at the University of California, Riverside.

Subsistence time. In contrast, subsistence time has greatly decreased, especially in the last few decades. Subsistence time is that time spent making a living or preparing to make a living. Not only has the total number of work hours per week declined, but the arrangement of work time has been influenced by such concepts as flex-time, work sharing, time-income tradeoffs, leaves without pay, sabbatical leaves, and graduated retirement (15).

Leisure/discretionary time. As subsistence time has decreased and changed, leisure time has increased. Leisure time is defined as spare time we are free to use for rest or to do with what we choose. The growth and size of the turfgrass industry is closely associated with the availability of true leisure or discretionary time for a large percentage of the population.

Discretionary income. In addition to discretionary time, discretionary income has increased substantially for the average citizen. Discretionary income—money left after necessary expenses have been paid—results from real growth in salaries, increases in two-income households, or the instituting of social programs based on income redistribution. The result has been the availability of more money for spending in a discretionary manner. Spending this money for recreational pursuits or beautification of property has had a large impact on the turfgrass industry.

Desire for amenities. With urbanization and increased discretionary time and money, a general desire for positive surroundings has evolved. This perception has been defined as "Those stimuli which lead to feelings of comfort, pleasure, or joy may be referred to as amenities" (3). Regarding amenities, subenvironments are identified within the urban-metropolitan environment as:

Residential environment:	Individual homes and surrounding immediate area.
Occupational environment:	Both interior and exterior of work/learn place.
Service environment:	The characteristics of institutions, organizations, and commercial establishments.
Leisure or recreational environment:	Places, facilities, and areas where urban residents go for rest within and without the urban complex.
Commuting environment:	The pathways that are used for transportation within and without the complex.

Turfgrass within the total planned landscape is a significant aspect of each subenvironment. Used wisely, it improves the subenvironment because of its functional and amenity value.

This concept has not gone unnoticed by planners and developers. New communities are designed with the understanding that attractive open space and recreational facilities are important aspects for resident satisfaction (and initial residence purchase). With such design and development, residents perceive that the quality of life is heightened by their community setting (5).

Population increase. The amount of turfgrass acreage apparently coincides with population size. As the U.S. population continues to increase, so, too, will the amount of turfgrass acreage.

Functions of turf

Many recreational facilities depend on a uniform, well-maintained turf sward as a medium of play. Common examples include golf courses; bowling greens; picnic areas and parks; soccer, lacrosse, polo, baseball, and football fields; and school grounds. Turfgrass provides the smooth surface required for many of these recreational activities and sites and also provides a safety "cushion."

Additionally, turfgrass, along with trees, shrubs, groundcovers, and flowers, is an important part of public and private gardens. Because gardening is a popular leisure-time activity, lawn maintenance can be a constant source of challenge and pride for those who enjoy this activity. And, finally, turf affects peoples' lives when used in an ornamental setting to create the desired aesthetic appearance.

Turfgrass not only influences our lifestyles but our environment. Turfs and other plant material reduce discomforting glare, especially in urban areas with buildings, metal, and concrete. Likewise, turfgrass, along with properly placed trees and shrubs, can considerably reduce traffic noise. Soil erosion is reduced or controlled by turf, and chemical and particulate air pollution is decreased at the turfgrass surface. Because of transpirational cooling, turf modifies high temperature by heat dissipation. (The obviously different temperature felt when standing on an asphalt pavement in comparison to a nearby turfed site is recognized by all.)

Certainly turfgrass is an important, positive modifier of our environment, and the turfgrass industry is an obvious source of economic activity. Recreational facilities and general lawns require regular maintenance, a sizeable category of economic activity. A second category of economic involvement is manufacturing—the production of equipment, fertilizer, chemicals, seed, sod, and other supplies. A service category includes individuals, groups, or firms—distributors, architects, contractors, and consultants—who provide services for the facility and manufacturing categories. Lastly, an institutional category involves those who conduct research and education to support the industry.

To view the impact of the turfgrass industry on a local area in terms of acreage and human involvement, turf in a hypothetical city was identified.

Turf in a hypothetical city

Watertown, USA, is a hypothetical city. Its population of 170,000 is located in one of the urban regions of the U.S. The city is located on major transportation arteries and is considered an important commercial, industrial, educational, and cultural center. Most residents work in or near the city.

In those positions directly involved with turf facility maintenance, 166 individuals are employed (Table 1, page 10). An additional 110 are employed in other sectors of economic activity supporting turf facilities. Landscape architects, landscape contractors, and equipment and material sales and service personnel depend, in varying degrees, on local turf facilities.

The largest category of turf installation (2) includes single-family and multiple-family residences, accounting for a total of 4,482 acres (Table 1). Single-family dwellings occupy 3,495 acres. Although relatively few are employed in the direct care of home lawns, those who work for nurseries, department stores, etc., earn incomes largely because of home lawn care.

Bowling greens used for lawn bowling. There are three greens in Watertown, USA. Cities with higher percentages of retired persons may have more greens, as bowling is popular with senior citizens.

A city cemetery. Three cemeteries have 90 acres of turf requiring maintenance by eight employees.

A city church: seventeen acres of turf surrounding 195 churches.

A general use park. The 628 acres of turf in the city's 50 parks require part- or full-time maintenance by 82 individuals.

A city golf course. The 599 acres of turf on six courses call for a work force of 40.

School turf site. The 56 public schools (K–12) have 400 acres of turfgrass maintained by 21 individuals.

Turf at factory site. The 350 factories are beautified by 47 acres of turf.

Residential lawn. An estimated 3,495 acres of turf are associated with single-family dwellings and 987 acres with multiple-family dwellings in Watertown, USA.

all photos this page by Max Clover

TABLE 1 Turfgrass in Watertown, USA, by category of use, size, and number of maintenance workers required

Turfgrass category	Number of facilities	Total acreage	Number of maintenance workers
Bowling greens	3	1	Under parks
Cemeteries	3	90	8
Churches	195	17	*
City parks	50	628	82
Golf courses	6	599	40
Colleges	2	156	15
Schools (K-12)	56	400	21
Factories	350	47	*
Residences:			
Single-family	45,200	3,495	*
Multiple-family	19,600	987	*
Total	65,465	6,419	166

*Maintained by volunteers or contracted individuals and firms.

In Watertown, the largest block of turf, besides home lawns, is the 628 acres of city parks. Golf courses use 599 acres of turf, schools have 400 acres, and colleges have 156 acres. There are 90 acres of turf in cemeteries, 47 acres in factory lawns, 17 acres for church lawns, and an acre of bowling greens. All told, Watertown has 6,419 acres of turf to maintain.

Average lawn sizes in Watertown are 750 square feet in the front yard and 1,000 square feet in the backyard. In caring for the average lawn, the homeowner spends $200 per year on fertilizer, pesticides, seed, and water. With 45,200 single-family dwellings, $9,040,000 per year is spent. The lawns associated with multiple-family dwellings are usually cared for by managers at the rate of $20 per unit. With 19,600 units in Watertown, $392,000 is spent on maintenance, including water.

City parks, cemeteries, schools, churches, colleges, and factories in Watertown represent 1,338 acres and spend more on water than any other item. Labor, supplies, equipment, and water cost $1,100 per acre, for a total of $1,471,800. Golf courses and bowling greens cost $2,200 per acre to maintain, for a total of $1,320,000 for 600 acres.

In Watertown, USA, $12,223,800 in direct expenditures on turf alone significantly affects the economy of the city. Multiplier effects, obviously, extend that manyfold. Turf is important to Watertown aesthetically, economically, and socially.

Watertown, with its 6,400 acres of turf and 166 individuals employed to care for them, is typical. However, for some parts of the U.S. this may be conservative. For example, in many areas, highway turf is an important category and accounts for many maintained turf acres. Certain cities have large turf areas associated with military complexes and regional, state, or federal parks. Some cities have a large concentration of sod farms. In general, cities such as Watertown, located where low or seasonal rainfall makes irrigation necessary, will have less turf and smaller turf plantings than areas with higher rainfall evenly distributed throughout the year. Nevertheless, Watertown illustrates the sizeable categories of turf, the number of acres, and the considerable employment opportunities and resultant economic activity associated with turfgrass in modern society.

The national turfgrass industry

Although turf was commercially recognized before World War II, the real growth and development of the turf industry occurred after the war. In 1965, it was reported (22) that turfgrass was a $4.3 billion per year industry. By 1982, it had grown to a nearly $25 billion industry. The fixed asset value of turf is, of course, many times that annual expenditure.

California, Florida, Michigan, New York, and Pennsylvania all have billion-dollar turf industries, and Illinois and Texas are very near this level (Table 2). As shown in the table, the turf industry has a significant economic impact on each state. (See Appendix A, *Background for cost figures in Table 2.*)

Two-thirds of all turf expenditures go to maintain home lawns. Much of this is retail sales, but the use of professional lawn care services is growing. The other categories utilize turf maintenance personnel with varying degrees of expertise. The "general lawns" category includes industrial parks, commercial buildings and developments, government buildings, and other nonresidential lawn areas.

Every mowed piece of turf represents an ongoing expenditure. In many areas, even unmowed turfgrass costs money in the form of water, labor, and, sometimes, fertilizer. To take care of turf, money is spent on seed, sod, sprigs, water, lawnmowers, rototillers, fertilizers, weed killers, a vast array of other products, and labor.

Even though the technology of turfgrass management has undergone tremendous development, it is still labor intensive. It is estimated that 380,000 people make their livings directly from the care and maintenance of turf in the United States (Table 3, page 12). If one were to consider the number that are indirectly involved—in manufacturing plants, dealerships, etc.—turf contributes significantly to employment.

Conclusions

The tremendous growth and resulting size of the turfgrass industry are due to our population's societal structure which requires turf and other amenities to modify and enhance our chosen living environments and to provide us with a lifestyle that we want. Fortunately, our discretionary income has been able to support activities and desires that our discretionary leisure time has made possible. Turfgrass in our society is an industry that generates economic activity in excess of $24 billion, which in itself demonstrates its importance.

The future of the turfgrass industry will continue to have a strong impact on our lifestyles and environment because our societal structure is not expected to change dramatically. This does not mean, however, that the industry can stagnate or continue to do "things as usual." The energy crises of 1973–1974 and the reduction of tax-supported programs of the late 1970s and early 1980s has placed financial restrictions on turf maintenance budgets in many facilities. The questionable availability of sufficient water of adequate quality and price in the future also will pose a challenge to the industry. The need for the industry to do the same job with fewer resources may be the thrust of the future.

TABLE 2 Annual turfgrass maintenance cost for eight turfgrass categories in 50 states (1982)

State	Home lawns	Golf courses	Parks	Schools	General lawns	Cemeteries	Highways	Airports	State total	Sod sales
					in millions of dollars					
Ala.	197.6	15.6	1.3	6.7	34.3	5.5	10.8	6.4	278.2	4.0
Alaska	15.1	.5	.9	.5	2.5	.4	.8	.5	21.2	0.0
Ariz.	350.6	21.6	22.4	11.7	60.1	9.7	18.9	11.2	506.2	2.4
Ark.	169.2	11.9	10.9	5.7	29.3	4.7	9.2	5.5	246.4	3.1
Calif.	648.9	124.2	47.7	24.9	128.0	20.6	40.1	23.9	1,058.3	30.4
Colo.	477.0	9.9	30.0	15.7	80.3	12.9	25.2	15.0	666.0	28.9
Conn.	467.5	39.4	30.5	15.9	81.8	13.2	25.6	15.2	689.1	4.4
Del.	40.0	5.6	2.7	1.4	7.4	1.2	2.3	1.4	62.0	.4
D.C.	40.0	1.7	2.5	1.3	6.7	1.1	2.1	1.2	56.6	0.0
Fla.	723.2	75.1	49.4	25.8	132.5	21.3	41.6	24.7	1,093.6	50.0
Ga.	331.7	45.2	22.9	12.0	61.4	9.9	19.2	11.4	513.7	8.7
Hawaii	200.7	16.5	13.0	6.8	34.9	5.6	10.9	6.5	294.9	0.0
Idaho	15.1	4.3	1.2	.6	3.3	.5	1.0	.6	26.6	1.8
Ill.	600.1	99.5	42.7	22.3	114.5	18.4	35.9	21.3	954.7	15.9
Ind.	552.8	37.4	35.5	18.6	64.6	15.3	28.3	16.2	768.7	6.2
Iowa	467.5	37.7	30.4	15.9	62.0	13.0	20.7	15.2	662.4	3.3
Kans.	458.1	29.2	29.4	15.3	78.7	12.7	24.7	14.7	662.8	4.4
Ky.	297.6	24.5	19.3	10.1	51.8	8.3	16.3	9.7	437.6	1.9
La.	303.2	11.5	18.9	9.9	50.7	8.2	15.9	9.5	427.8	2.6
Maine	20.1	14.9	2.1	1.1	5.6	.9	1.8	1.0	47.5	0.0
Md.	514.9	18.3	32.2	16.8	86.3	13.9	27.1	16.1	725.6	12.8
Mass.	486.5	36.9	31.4	16.4	84.2	13.6	26.4	15.7	711.1	.4
Mich.	675.9	60.8	44.6	23.3	119.7	19.3	37.5	22.3	1,003.4	17.9
Minn.	505.4	35.3	32.9	17.2	88.3	14.2	27.7	16.5	737.5	16.8
Miss.	269.2	11.2	16.9	8.8	45.4	7.3	14.2	8.5	381.5	5.1
Mo.	325.0	29.4	21.3	11.2	57.3	9.2	18.0	10.7	482.1	8.3
Mont.	15.1	4.1	1.2	.6	3.1	.5	1.0	.6	26.2	.8
Nebr.	166.6	22.6	10.4	5.5	28.0	4.5	8.8	5.2	251.6	4.3
Nev.	60.0	7.6	4.0	2.1	10.9	1.7	3.4	2.0	91.7	0.0
N.H.	101.5	10.5	6.7	3.5	18.0	2.9	5.6	3.4	152.1	0.0
N.J.	514.9	50.7	34.2	17.9	91.9	14.8	28.8	17.1	770.3	17.3
N.Mex.	107.0	6.8	6.9	3.6	18.5	3.0	5.8	3.4	155.0	1.9
N.Y.	732.7	130.4	52.0	27.2	139.5	22.5	43.7	26.0	1,174.0	13.2
N.C.	337.3	39.2	22.6	11.8	60.6	9.8	19.0	11.3	511.6	.9
N.Dak.	19.5	14.9	2.1	1.1	5.5	.9	1.7	1.0	46.7	0.0
Ohio	590.6	63.8	39.5	20.7	106.0	17.1	33.2	19.8	890.7	9.5
Okla.	357.0	13.1	22.3	11.6	59.8	9.6	18.7	11.1	503.2	3.1
Oreg.	467.5	10.6	28.9	15.1	77.4	12.5	24.3	14.4	650.7	4.8
Pa.	675.9	66.8	44.6	23.3	119.7	19.3	37.5	22.3	1,009.4	5.2
R.I.	200.7	7.4	12.6	6.6	33.9	5.5	10.6	6.3	283.6	9.0
S.C.	280.5	21.7	18.1	9.5	48.6	7.8	15.2	9.1	410.5	.3
S.Dak.	19.5	15.1	2.1	1.1	5.6	.9	1.8	1.0	47.1	.7
Tenn.	314.6	24.6	20.3	10.6	54.6	8.8	17.1	10.2	460.8	1.1
Tex.	647.5	40.6	41.8	21.8	112.0	18.0	35.1	20.9	937.7	31.0
Utah	41.7	4.6	2.9	1.5	7.7	1.2	2.4	1.4	63.4	3.0
Vt.	19.5	7.8	1.6	.9	4.4	.7	1.4	.8	37.1	0.0
Va.	552.8	26.9	34.8	18.2	93.4	15.0	29.3	17.4	787.8	4.5
Wash.	514.9	15.1	32.0	16.7	85.8	13.8	26.9	16.0	721.2	6.1
W.Va.	163.5	15.8	10.8	5.6	28.8	9.0	9.0	5.4	247.9	.1
Wis.	533.8	39.5	34.9	18.2	93.5	15.1	29.3	17.4	781.7	12.1
Wyo.	15.1	3.2	1.1	.6	2.9	.5	.9	.6	24.9	1.0
Total	$16,602.6	$1,481.5	$1,081.4	$571.2	$2,881.7	$476.3	$912.7	$545.0	$24,552.4	$359.6
Employed directly	54,200	38,700	42,400	28,000	113,000	23,300	44,700	26,700	371,000	9,400

Total employed   380,400

TABLE 3 Number of persons employed directly in the U.S. turfgrass industry

Category	Total expenditures (× $1,000)*	Labor (%)†	Employed‡
Home lawn	16,602.6	5	54,200
Golf courses	1,481.8	40	38,700
Parks	1,081.4	60	42,400
Schools	571.2	75	28,000
General lawns	2,882.7	60	113,000
Cemeteries	476.3	75	23,300
Highways	912.7	75	44,700
Airports	545.0	75	26,700
Sod farms	359.6	40	9,400

*Total expenditures are found in *Table 2*.
†Estimated.
‡Total expenditures × labor % divided by $15,300. Average work hours/year (1,800) × Average $/work hour ($8.50) = $15,300.

Appendix A. Background for cost figures in Table 2

Several states have surveyed their own turfgrass industries (1, 4, 11, 12, 14, 16, 17, 23, 24, 26), indicating actual growth well beyond inflation. The effects of inflation were then computed (10, 21).

To estimate home lawn valuation, a linear regression was calculated, using home lawn data from six surveys (11, 14, 16, 23, 24, 26). These data were plotted against Occupied Housing Units from the Census of Housing (8, 9). The surveys were updated using the Consumer Price Index (10).

The regression was computed, using the adjusted values, as $Y_c = 372.9 + 94.69X$, where Y_c is the computed home lawn value and X is the Occupied Housing Units. The correlation coefficient is 0.6024. Adjustments then were made for states with high central city population density, low population density (rural), populations under 1 million, low per capita income, and small lawns in arid irrigated areas with high water costs.

The National Golf Foundation published the number of golf facilities in each state (13). Using data from the Golf Course Superintendents Association (18), the Hawaii survey (25), and the CPI, the numbers of facilities and costs were adjusted to 1982.

Sod farm sales, in acres, were taken directly from the Census of Agriculture (6, 7). Sales dollar values per acre, regionally, were obtained from members of the American Sod Producers Association. Alabama had a recent in-state survey of the sod industry (1) and that value was used without adjustment.

The other categories were computed as percentages of the means from the surveys. These percentages then were applied to the previously calculated values.

Literature cited

1. ADRIAN, J. L., J. A. YATES, and R. DICKENS
   1981. Commercial Turfgrass-Sod Production in Alabama. Alabama Agricultural Experiment Station. Auburn Univ. Bull. 529.
2. ANONYMOUS.
   1981. Community Economic Profile. Riverside County Department of Development. 185–189.
3. ATKISSON, A., and I. ROBINSON
   1979. Amenity Resources for Urban Living. Chapter 20. *In* Van Doren, C., G. Priddle, and J. Lewis. Land and Desire: Concepts and Methods in Outdoor Recreation, 2nd edition. 317 pp.
4. BEUTEL, J., and F. ROEWEKAMP
   1954. Turfgrass Survey of Los Angeles County. The Southern California Golf Association.
5. BORBY, R.
   1975. Recreation and Leisure in New Communities. Ballinger Publishing Company, Cambridge. 366 pp.
6. CENSUS OF AGRICULTURE
   1970. U.S. Department of Agriculture.
7. ———
   1978. U.S. Department of Agriculture.
8. CENSUS OF HOUSING: GENERAL HOUSING CHARACTERISTICS
   1970. U.S. Department of Commerce, Bureau of the Census.
9. ———
   1980. U.S. Department of Commerce, Bureau of the Census.
10. CONSUMER PRICE INDEX DETAILED REPORT TO OCTOBER 1982
    1982. U.S. Department of Labor, Bureau of Labor Statistics.
11. FLORIDA TURFGRASS SURVEY
    1976. Florida Crop and Livestock Reporting Service.
12. GIBEAULT, V.
    1979. Importance of Turfgrass in California. California Turfgrass Culture 29(4):25–26.
13. GOLF FACILITIES IN THE U.S.
    1972. National Golf Foundation. ST-1.
14. GRUTTANDAURIO, J., E. E. HARDY, and A. S. LIEBERMAN
    1978. Abbreviated Report on: An Investigation of Turfgrass Land Use Acreages and Selected Maintenance Expenditures Across New York State. New York State College of Agriculture and Life Sciences.
15. KONIGSBERG, D.
    1978. Laboring Over the Work Crisis. *In* American Way. American Airlines. Sciences 11(12):47,50.
16. MICHIGAN'S $350 MILLION CARPET
    1969. Michigan Science in Action. Michigan State Univ., Agricultural Experiment Station.
17. NEW MEXICO TURFGRASS SURVEY
    1980. New Mexico Department of Agriculture.
18. PRUSA, J.
    US GCSA, personal communication November 1981.
19. SCHERY, R.
    1961. The Lawn Book. The Macmillan Company, N.Y. 207 pp.
20. STATISTICAL ABSTRACTS OF THE U.S.
    1981. National Data Book and Guide to Sources. U.S. Department of Commerce, Bureau of the Census.
21. TOFFLER, A.
    1980. The Third Wave. Bantam Books, N.Y. 537 pp.
22. TURFGRASS AS A $4 BILLION INDUSTRY
    1965. Turfgrass Times 1:1.
23. TURFGRASS MAINTENANCE COSTS IN TEXAS
    1964. Texas Agricultural Experiment Station.
24. TURFGRASS SURVEY
    1966. Crop Reporting Service. Pennsylvania Department of Agriculture.
25. VAN DAM, J., and C. L. MURDOCH
    1976. Turfgrass Maintenance on the Island of Oahu. Hawaii Agricultural Experiment Station, Univ. of Hawaii, DP 41.
26. WASHINGTON TURFGRASS SURVEY
    1967. Cooperative Extension Service, Washington State Univ.

2 Water: Whose is it and who gets it?

Joseph P. Rossillon

JOSEPH P. ROSSILLON

Joseph P. Rossillon holds a Ph.D. in speech and communications with an emphasis on the psychology of communication. Currently, he serves as president of the Freshwater Foundation, an organization in Minnesota that generates and distributes funds to support research on freshwater, and information and education programs designed to translate water issues and scientific information for the general public.

Across the United States, water-related problems are increasingly leading the news. As we view the different, apparently isolated problems of this nation, it is important that they be viewed not as isolated problems but as local manifestations of a total system under stress.

It is important for us to understand that our many water problems are, in actuality, only one water problem: We are overusing a natural resource. The answer to the question, "How did we get in this fix and what can we do?" begins with reviewing these questions:

- How much water is there?
- Who owns water?
- Who manages water?
- Who uses water?
- What are the priorities?
- What are the options?

How much water is there?

Water, like energy, can neither be created nor destroyed. When we speak of "more or less" water, we speak not of the total amount available, but of the amount of available, usable, affordable water at the right place and at the right time. In this respect, the current water situation in the United States is similar to the energy situation we faced in 1973. Those similarities should be recognized.

Like everything else, energy is controlled by laws of nature, including two laws of thermodynamics: One law deals with the constant, the other with change. The first law says energy can neither be created nor destroyed. The second law says the universe is constantly and irreversibly becoming less ordered than it was. We choose to believe in the first law and forget the second. In the process we fail to differentiate between the terms quantity and available in a usable condition, which at any given time can deviate significantly.

In the United States, before 1973, the energy supply was either sufficient or we had the "power" to make it available. In the mid-1970s we lost control over the energy sources and were unable to keep an adequate supply available; thus, the energy crisis.

Similar rules affect water. There are two laws of aquadynamics: The first law (quantity) says, "Water can neither be created nor destroyed"; the second law (quality) says, "Through its use, water is constantly and irreversibly changed."

When we address the question of water shortages, we refer not to the matter of "enough water" but rather to the issue of enough usable water at the right place, at the right time, of an acceptable quality, and at a legitimate and acceptable price. In other words, we refer to "available, usable water."

Of all the water in our universe, the available, usable portion represents less than 1 percent of the total amount. About 97 percent is salt water, and the other 2 percent is polar ice. That 1 percent has been more than sufficient to meet our needs in the past—but no longer.

The hydrologic cycle that replenishes the water supply on our earth is like a huge recirculating pump. The natural process has been replacing more water in our available, usable reservoirs than has been needed to meet our daily requirements. We have been able to draw from "surplus" supplies—but no longer. Now drawdown exceeds the recharge in many areas of the United States.

The increase in drawdown can be traced primarily to (a) increased population, especially with population shifts to low-water-volume areas like the Sun Belt; (b) obsolete technology—technology that is intensive in its use or that is wasteful because of design or obsolescence; and (c) intensive irrigation farming to produce "high-water-use" crops.

A different factor that affects the amount of water available for use is the loss of "usable" water supplies. This phenomenon occurs with the saline recharge of coastal groundwaters, with the contamination of major groundwater areas by toxic chemicals, with collapse of major aquifers as occurred in the San Joaquin Valley, or with the drawdown of glacially filled nonrechargeable aquifers like portions of the Ogallala. (See chapter 3, *Water resources in the United States*.)

As a result of this series of situations, we're beginning to use more water annually in the United States than is available. This shortage of available water necessitates creation of priorities and increases water's value.

Who owns the water?

Historically, individual states have owned the water "above, on, and beneath the land of the state." As an owner, each state established and enforced its own water law. Colonists in the East brought with them English water law, the doctrine of riparian rights. As described by Dr. William A. Thomas, consultant to the American Bar Association,

> A riparian owner is one who owns land abutting to a stream or lake. Under the doctrine's early formulation, ownership of riparian land carried with it a property right to have the water course maintained in its natural condition. All riparian owners had equal rights to use the water but only in those ways that would maintain the stream in its natural condition, both as to quantity and quality of water.[1]

Water law in the western United States developed differently. Mining districts, to preserve their rights to water flowing through their areas, devised the doctrine of prior appropriation. According to Dr. Thomas,

> Under this doctrine the first person to divert water from a water course and put it to beneficial use thereby receives a right in perpetuity to use that amount of water as long as it is put to beneficial use. This "first in time, first in right" arrangement means that no one can divert water until all persons with prior rights have had an opportunity to divert the water to which they are entitled.[2]

Viewed together, these separate governing laws are in direct conflict. The doctrine of riparian rights is based upon equality of rights; under the prior appropriation doctrine, different uses of water can never have equal rights.

Under the prior appropriation law, no one is technically responsible for maintaining the natural condition of the water, and the accelerated use of groundwater under these conditions raises the problem of "mining of water." If the volume of water removed from an aquifer exceeds the amount recharged to it, the overdraft eventually will exhaust the water supply. This phenomenon accounts for the drawdown of the aquifer of the San Joaquin Valley, the lowering of the aquifer basins near Houston, Texas, and the significant drawdown of the Ogallala Aquifer in the Midwest.

Adding to the complexity of who is responsible for water is the newly addressed question of who owns it. A federal judge ruled that the city of El Paso, Texas, can drill 326 wells in neighboring southern New Mexico and import up to 100 billion gallons of water per year to help offset a water shortage created by its rapid growth. The decision was based on a Supreme Court decision rejecting the legal doctrine of state ownership of water. Interpreting the Supreme Court's decision in Sporhase vs. Nebraska, the lower court stated that "a state may discriminate in favor of its citizens only to the extent that water is essential to human survival. Outside of fulfilling human survival needs, water is an economic resource. For the purpose of constitutional analysis under the common clause, it is to be treated the same as other natural resources."[3]

In summary, it appears that a state can own its water resources as economic and natural resources, but not to the detriment of the needs of the human population. This decision can be viewed as the beginning of the establishment of a priority system for ownership and usage of water.

Minnesota Statute 105.41 in 1978 established a priority listing for water use. It stated that rules governing water would be based on the following priorities for appropriation and use: First priority, domestic water supply excluding industrial and commercial uses of municipal supply; second priority, any use involving less than 10,000 gallons of water per day; third priority, agricultural irrigation; fourth priority, power production; fifth priority, other.[4] This statute was tested by the Minnesota Supreme Court in December 1980 and reheard in January 1981 in a conflict between the city of Crookston and Crookston Cattle Company. The statute was held constitutional.

In a third test of ownership of water, South Dakota, in 1981, agreed to sell Energy Transportations Systems, Inc. water from behind Oahe Reservoir of the Missouri River to build a coal slurry pipeline.[5] The question of whether the state had the right to sell river water behind a federal impoundment is presently being asked in the courts by the states of Iowa and Nebraska.

Who manages water?

Perhaps as confusing as the question, "Who owns water?," are the questions, "Who manages it and how is it managed?" Ac-

---

[1] William A. Thomas, "Allocation of Water as a Scarce Resource," *Futurics*, 1981.
[2] *Ibid.*
[3] El Paso v. S. E. Reynolds *et al.*
[4] Minnesota Statute 105.41 (1978).
[5] "ETSI Buys Missouri River Water," *Upper Midwest Report*, September–October 1981.

cording to *Newsweek*, "One big obstacle to good infrastructure maintenance of water is the very system that controls it, i.e., 100 federal agencies, 50 states, 3,000 counties, and thousands of local agencies."[6] In April 1980, the U.S. Water Resources Council published a report, "State of the States: Water Resources Planning and Management." The council's task was to examine state statutes, policies, regulations, and administrative mechanisms dealing with water resource management and planning. The report presents some interesting, if not startling, observations. For example:

> Twenty-one states do not have the express legislative or administrative mandates to undertake comprehensive water resource planning. Only two states, Florida and Delaware, have fully comprehensive approaches with planning and management functions consolidated in one agency with authorization to conduct comprehensive, multipurpose planning. Of the remaining 27 states, 11 have comprehensive quality and quantity planning (but not management) integrated in a single agency; six have quantity (but not quality) planning and management functions integrated in one agency; 10 have agencies authorized to undertake water quantity planning (but not quality planning or management functions).[7]

The study concludes, "It is common to find water quality and water quantity administered by separate units of government, each exercising authority in only one field, without any effective legislative authority to bridge the gap."

This lack of coordinated management and planning may explain, in part, why the limited available, usable water resources are being overtaxed and why "weak-link stress points" increasingly occur across our United States.

Current management policies and practices are structured around the question: "Who owns the water?" As supplies of available water decrease and the priorities for use become significant issues, emphasis will shift from "Who owns it?" to "Who gets to use it?" Management structures will need to be identified to reflect better how water is used and who uses it, if we are to adequately protect highly stressed limited use resources.

Who uses water?

It is important to distinguish between use and consumption of water. Much of the confusion over "who uses the most" comes from a lack of clarification between use and consumption. Writer Stephen D. Morton presents the following definition and distinction:

> When we speak of water being "used," it is generally meant that some impurity is added to the water, and the water is then put back into some river, lake, or ocean. This is called nonconsumptive use since the water is only dirtied. It does not actually disappear. Consumptive use is when the used water seems to disappear; that is, the water is no longer available in the area where it came from. Irrigation is the main consumptive use of water. Irrigation is most widely used where it is dry and hot; therefore, a large amount of water evaporates rather than becoming surface or groundwater. Some of the water is also incorporated into the crops by photosynthesis. This distinction between consumptive and nonconsumptive use is more than just a minor distinction. It is very important when one considers how much more water can in effect be made available by cleaning polluted water and reusing it over and over again. Obviously only water used nonconsumptively can be reused in this way.[8]

Considering all of our uses for water, the average per-person water use in the United States is between 1,800 and 2,000 gallons per day. Industry accounts for 43 percent of the use, irrigation for 47 percent, and domestic use for cooking, bathing, sanitation, drinking, lawn sprinkling, etc. for the remaining 10 percent. These use percentages were best put into perspective in an article in *Forbes*, which concluded, "Bricks in toilet tanks or shutting off sprinklers hissing on summer lawns makes better symbolism than sense in dealing with water shortages."[9] A chart from G. Tyler Miller, Jr.[10] best puts the use of water by Americans into perspective.

An American's insatiable thirst: The average American uses 1,800 gallons of water per day; the average person in an underdeveloped country uses 12 gallons per day.

Direct personal use: 160 gallons per day—8 percent of our daily use

Bath: 30 to 40 gallons

Shower: 5 gallons per minute

Shaving (water on): 3 gallons

Washing clothes: 20 to 30 gallons

Cooking: 8 gallons

Washing dishes: 10 gallons

Flushing toilet: 3 gallons (110 gallons per day for a family of four)

Leak in toilet bowl: 35 gallons per day

Scrubbing and cleaning house: 8 gallons

Sprinkling 8,000-square-foot lawn: 80 gallons

Faucet and toilet leaks in New York City: 200,000,000 gallons per day

Indirect agricultural use: 600 gallons per day—33 percent of our daily use

One egg: 40 gallons

One ear of corn: 80 gallons

One loaf of bread: 150 gallons

One gallon of whisky: 230 gallons

Five pounds of flour: 375 gallons

One pound of beef: 2,500 gallons

---

[6] "The Decaying of America," *Newsweek*, August 2, 1982.
[7] *Water Resource Management: New Responsibilities for State Governments*, Lexington, Kentucky, 1981.
[8] Stephen D. Morton, *Water Pollution—Causes and Cures*, Madison, Wisconsin, 1976.
[9] Kathleen K. Wiegner, "The Water Crisis: It's Almost Here," *Forbes*, August 20, 1979.
[10] G. Tyler Miller, Jr. *Living in the Environment*, Belmont, California, 1979.

Indirect industrial use: 1,040 gallons per day—59 percent of our daily use

Cooling water for electrical power plants: 720 gallons per person per day

Sunday paper: 280 gallons

One pound synthetic rubber: 300 gallons

One pound aluminum: 1,000 gallons

One pound steel: 35 gallons

One gallon gasoline: 7 to 25 gallons

One automobile: 100,000 gallons

According to author Fred Powledge, one calculation by the University of California in 1979 was that 4,533 gallons of water are required to produce the food that a Californian eats in one day, "assuming that the eater consumed 2,500 calories."[11]

According to the St. Regis Paper Company, "A major paper mill uses between 15 and 40 million gallons of water daily."[12] That totals 14 billion gallons a year. St. Regis has 13 major paper mills in the United States and estimates that it uses approximately 162 million gallons of water a day to produce its product.

According to Raymond Stack, New York City's water system probably loses 100 million gallons of water a day. Boston loses 2 gallons for each gallon delivered. Broken water mains in the New Jersey area spill more than 15 million gallons of water a year.[13] These figures demonstrate the desperate need for repair and replacement of antiquated water systems, but these systems are in cities already in financial crisis.

The operation of our society today is based upon cheap, safe, usable, readily accessible water supplies. Those supplies are diminishing. As our demands for water use increase, pressures are being placed upon the right of ownership, on priorities of use and on a need to look at new and more efficient technology and use practices. As the supplies of available water diminish, the question becomes not "Who uses it?" but "Who gets to use it first?" Shortages establish priorities and priorities initiate increases in value.

What are the priorities?

There was a time when there was plenty of water for everyone. Since that time we have begun to locate major population pockets where there isn't an adequate water supply and where the population has increased. The water available for use is the same amount. We now have more people using it and we have more people in places where there isn't enough water available for use. As a result, we no longer have the luxury of saying, "There's plenty of water for everyone and everything." Thus appears the issue of priorities.

Before we begin looking for new answers, let us begin by establishing some of the things we do know. We know, for example, that there is no "new" water. We have the same stuff we have always had and we are using it over and over again. The total volume of usable fresh water is less than 2,000 gallons per day per man, woman, and child in the world. In the United States, we are nearing a daily use in excess of 2,000 gallons per day per person which means we are beginning to use more than our fair share of the global fresh water supply. That is why pressures are occurring on our supply that previously didn't exist.

It is important to understand that there are no new supplies of water. One error we make when we consider redistribution of water supplies is the masking of our problems. Redistribution gives the impression that some new supply of water has been found. It hasn't; it has just been moved. It is part of some other supply that has been relocated.

A limited supply of water to meet our needs leads to the question, "Who gets to use it first?" Current and pending legislation based upon court cases suggest the people and the needs of people will always receive first priority. In other words, the first priority will always be potable water.

When referring to potable water, the issue of quality must be addressed concurrently with the issue of quantity. The World Health Organization states that 70 percent of the people in the world do not have acceptable drinking water, and that percentage is going up. The Environmental Protection Agency states that 40 percent of the municipal water supplies in the United States do not meet minimal standards. A study by Cornell University points out that 60 percent of rural water supplies in the United States are contaminated.

William Marks of the Water Management Planning Task Force of Michigan has stated, "I'm not about to say that the groundwater for the state of Michigan is totally contaminated, but I'm willing to predict that within five years the state's entire water supply will come from the Great Lakes."[14] The significance of that statement is not in the announcement of a new supply, but in the recognition of the loss of a former supply.

Contaminated water supplies in conjunction with population distribution in arid regions lead to a second priority issue: competition for limited supplies. In the arid Southwest and southern California, water supplies are already overextended. By compacts among seven states and a treaty with Mexico, the Colorado River is overbooked to 110 percent. A desalination plant is being constructed in Yuma, Arizona, to reclaim the water overstressed by irrigation. Most of the water assigned through this system is for agriculture. But as the population of the Sun Belt increases, orange groves and cotton fields are retreating, and housing developments are spreading. If the water supply is already overbooked, as the population continues to grow, the question will not be, "Where can we get more water?," but "Who will get to use it first?" The answer appears to be "the people."

If human consumption has the first priority and if there are no apparent new water supplies, a logical alternative would seem to be to recycle. This use priority might best be termed "think dirty."

Traditionally, our water has been supplied in a "single system," providing potable water for all uses. With a surplus of

---

[11] Fred Powledge, *Water*, New York, 1982.
[12] "As Essential to St. Regis as to Life Itself," *Reach*, Fall 1982.
[13] Raymond F. Stack, "Our National Water Crisis, Part One," *Boca Raton*, Winter 1983.

---

[14] William Marks, address to the Conference on Great Lakes Resources, Chicago, Illinois, March 4–5, 1982.

readily available cheap water, this has appeared to be the most convenient and economical system. As our supplies decrease, however, alternative systems may appear more popular and economically feasible. Irrigation systems, sprinkling systems, cooling systems, or almost any process that does not include human consumption might be a prime candidate for recycled or nonpotable water.

Examples of what could almost be termed nontraditional uses are already prevalent across the United States. In parts of Florida, housing developments depend upon golf courses to accept water from their sewage treatment plants. Since its inception, the Air Force Academy in Colorado has relied upon sewage water to maintain its ornamental and recreational areas. Closed-loop systems are being developed around the country, and surplus water from runoff is beginning to be incorporated as cooling water in some large plants.

The third priority issue will be the question of value or economics. When any item increases in priority, it subsequently increases in value. The economic structure of the United States is based upon the premise of free water. Historically, the only value placed upon water was for pumping, storage, cleaning, or delivery. The recent proposed sale of Missouri River water by the state of South Dakota may have changed all that. The value of that water at the source for sale purposes by South Dakota was priced at $.002 per gallon. If a similar charge for water is to be added to all final products using water, a trickle-up effect would occur on all products. Charging 2 cents for every 10 gallons of water used in production would, for example, increase the cost of an automobile by $200; a set of tires by $60; a bushel of corn by $10; hamburger, french fries, and cola by $3; and a pound of beef by $8. It would increase by $3,000 the cost of producing food for just one person for a year in the United States.

The potential impact of South Dakota's trendsetting proposal is significant. If any producer must add to production costs the cost of water at the source, the immediate questions to be asked would be: How badly do I need it? How much do I need? What type of water do I need? Will I be able to recoup the cost? What will be the impact on the buyer?

What are the options?

In looking for solutions we must first accept the reality that there are no easy answers. The easy solution or "quick fix" is always the first route used; however, it is usually applied to short-term, simple issues. Most of our easy-solution opportunities have already been used in dealing with water resources. What are some of the remaining options?

One option is interbasin transfer. The value of interbasin transfer of water is the ability to relocate water supplies proportionate to the redistribution of the population. This technique almost becomes mandatory if we are to provide our population with free access to relocation.

The disadvantage of interbasin transfer is that it tends to mask problems. Each time water is redistributed from one location to another, it tends to give the impression that a new source of water has been discovered. In fact, it is merely a relocation of an already overstressed supply.

The second option for relieving the pressures of limited water supplies is placed upon desalination. Vast oceans as potential water supplies are viewed by a thirsty nation as a hungry man views a banquet feast. Unfortunately, the current desalination technology is not adequate enough to provide major supplies of water. It may be possible to supplement the overstressed resources of coastal communities, but the volumes will be limited and the cost will be significant. Certainly the cost will be prohibitive for high water-use production such as is needed for irrigation farming.

A third option and a new potential supply of usable water can be obtained through cleanup of contaminated water supplies. As the state of Michigan plans to provide water for its population from the Great Lakes (because of contaminated groundwater supplies), the pressure on the Great Lakes will be increased. At the same time, a major supply of water in that aquifer will be sitting unused. New science and technology that will enable the reclamation of contaminated water will increase the available, usable water supply. Such science and technology will have long-lasting implications and benefits but will be slow and costly in coming.

The ultimate option is that we as a society must learn to live in harmony with those natural resources upon which we depend. The *Global 2000 Report to the President of the United States* concluded:

> The available evidence leaves no doubt that the world—including this nation—faces enormous, urgent, and complex problems in the decades immediately ahead. Prompt and vigorous changes in public policy around the world are needed to avoid or minimize these problems before they become unmanageable. Long lead times are required for effective action. If decisions are delayed until the problems become worse, options for effective action will be severely reduced.[15]

We must not think of our water as in a friend/adversary relationship: a friend when we are enjoying its benefits, an enemy when it floods, smells bad, tastes bad, or dries up. We must think of water instead as a constant indicator of how we as a society are doing. Our water is a constant measure of how healthy we are as a society. In other words, our water supplies are to our society what litmus paper is to a diabetic: They tell us how well our water system is in balance today. If the paper is bad, the diabetic doesn't change the paper; the patient's internal system is changed through medication or diet. So also with our water. If our water supplies are unacceptable, adjusting the supply will not solve the problem. We must find a way to make our water uses compatible with the supplies.

In analyzing our options and preparing for change, we must remember that human beings do not change water systems; they "intervene" into those systems. The question is not "What can we do with our water?" but "What will be the impact of our intervention?"

We must be careful about what we decide.

---
[15] *Global 2000 Report to the President: Entering the 21st Century*, Washington, D.C., 1980.

Practicum

3 Water resources in the United States

JAMES R. WATSON

James R. Watson, an agronomist, received his B.S. degree from Texas A & M University and his Ph.D. degree from Pennsylvania State University. He has been employed by The Toro Company at Minneapolis, Minnesota, since 1952, and is presently its vice president. In 1983 he was appointed an adjunct professor in the department of horticulture science and landscape architecture, University of Minnesota.

The United States has an abundant, currently adequate, and generally dependable supply of fresh water. Locally and regionally there are problems associated with the unequal distribution of rainfall that result in floods and drought. Other problems arise from underdeveloped raw water resources, overpumping of aquifers, legal restrictions on water rights to surface and underground sources, inadequate municipal distributive systems, waste, pollution of surface and underground waters, saltwater intrusion, an unrealistic cost structure and, perhaps, the bureaucratic inefficiency that results from lack of a single national controlling body for the innumerable independent and governmental agencies currently involved with the nation's water.

Water for turfgrass facilities is often limited or restricted because of one or more of the above problems.

Less than 1 percent of the world's water is available for man's use. Ninety-seven percent is tied up in the oceans and seas. Another 2 percent is tied up in glacial ice.

Nationwide water resources, in total, are more than adequate for current and foreseeable needs. Consumptive use in the conterminous 48 states (106 billion gallons daily) is only 7 percent of the total renewable supply (1,450 billion gallons daily). There are, also, extensive groundwater supplies, estimated to be between 33,000 and 59,000 trillion gallons. (For comparison, Lake Michigan contains 1,290 trillion gallons.)

In the contiguous 48 states, approximately 80 percent of the water resources are located in aquifers. They are the source of approximately 20 percent of the fresh water used nationally. This use is increasing at an annual rate of 3.8 percent.

Significant depletion of major aquifers has occurred in the Central Valley of California, southern Arizona (arid, semiarid areas), and the High Plains (subhumid Ogallala aquifer). Also, increased pumpage (from 100 *million* gallons per day in 1900 to approximately 1.5 *billion* gallons per day in 1980) has caused major declines in water levels of the Coastal Plain states—humid regions of high rainfall.

The factors that most often limit or restrict greater utilization of surplus runoff are:

- Insufficient surface storage areas,
- The extreme variability of annual precipitation that may result in floods or drought, and
- Legal and economic issues.

Turf cover reduces runoff and protects the soil surface from erosion, thereby helping to prevent accumulation of sediments in surface waters. Additionally, turf cover will filter atmospheric pollutants and particulate matter brought to earth by precipitation. This minimizes the adverse impact of "acid" rain in those areas affected by this phenomenon.

The nation's turfgrass facilities must not be overlooked in the development and execution of water policy. They contribute to the economy and recreational value of our communities; they are aesthetically appealing and functionally beneficial. They are a national treasure. Political entities and local authorities need to be made aware of the importance and economic value of green spaces to ensure a fair and just allocation of water resources.

The turfgrass industry must seek its fair allocation of water. Further, it must seek and develop alternative sources. Among those to be considered are wastewaters, including sewage effluents; capture and impoundment of runoff waters; and dual water systems for turf facilities, including home lawns, to accommodate potable and nonpotable waters.

The industry must conserve water, develop drought- and salinity-tolerant grasses, and devise cultural practices that reduce the amount of water used for turfgrass purposes.

3 Water resources in the United States

James R. Watson

Historically, much of the United States has treated water as if it were an unlimited and inexhaustible resource. Changing attitudes about our water resources relate to the availability of water, its location, its costs, the role of surface waters, and the pumping of underground reservoirs (20, 23, 24).

Water is a limited renewable resource; it is not always available in usable form in the amounts needed. Further, there is an ever-increasing demand for water. And, as uses increase, the concept that "water is free" must be abandoned. Water has really never been free for most purposes, but usually an actual or realistic cost has not been paid. For example, an estimated 40 gallons of water are used to produce one egg, 150 gallons to produce one loaf of bread, and 2,500 gallons to produce one pound of beef. If water were to cost only 1 cent per gallon, water cost for the egg would be $1.20, for the bread $3, and for the beef $40 (11, 27). Some are beginning to believe that a cost of a tenth of a cent per gallon of water is not unrealistic. What the cost per gallon may be 10, 20, or 50 years from now is unknown; however, it seems reasonable that as more demands are placed on available supplies, the price of water will escalate. It is clear also that cost will vary markedly with geographic location. The water-short Sun Belt areas with their exploding populations will be the first to feel the impact of inadequate water supplies and water at "unreasonable" costs. However, as the era of "free water" passes, costs will reflect more accurately water's true value and prices for food and other products requiring water for processing will become markedly higher.

There is increasing evidence that with planning, conservation, and management our limited water resources may provide for our needs well into the future. Contributing to this attitude are studies conducted by federal, state, and local agencies: the U.S. Water Resources Council, associated state agencies like the Water Resources Research Centers, and local underground water conservation districts like the High Plains Underground Water Conservation District Number 1, and The Freshwater Foundation and similar organizations (9, 28). The concepts stemming from these areas of investigation are supplanting earlier (perhaps misdirected) beliefs that development of surface waters and increased pumping of deep underground reservoirs will provide for all of our current and future water needs.

The work of these and other governmental agencies concerned with water resources (see list) continues to stimulate conservation efforts, reuse and recycling of waste waters, desalination, weather modification, and development of secondary recovery techniques from certain aquifers (6, 28). All of these water management practices will help to maintain, augment, and perhaps increase future supplies of usable water.

Required also are development of discretionary use of potable versus nonpotable water, and, in the turfgrass field, breeding, development, and widespread use of drought- and salinity-tolerant species and cultivars (32, 33).

As will be noted, the United States has an adequate supply of water from surface and underground sources, but at any time regional shortages of water may occur because of droughts and the uneven and untimely distribution of precipitation. Too, excessive amounts of rainfall and snow melt sometimes cause local

A partial listing of governmental and independent agencies concerned with water resources

FEDERAL LEVEL

Departments of: Agriculture, Defense, HUD, Interior, Commerce, Energy, and Transportation

U.S. Corps of Engineers (Department of Defense) and numerous district offices
Bureau of Reclamation (Department of Interior)
National Science Foundation
Water Resources Council

REGIONAL LEVEL

Environmental Protection Agencies
Interstate Commerce Commission
Various river basin commissions (21)
Great Lakes commissions
Various boundary water commissions
Resource conservation and development areas

STATE LEVEL

Departments of: Agriculture, Economic Development, Fish and Game, Public Safety, Health, Natural Resources, and Transportation

Environmental quality boards
Pollution control agencies
Soil and water conservation boards
State planning boards and agencies
Water planning boards
Water resources boards

LOCAL LEVEL

County agencies, boards, and committees
Municipal agencies
Township agencies
Drainage districts
Watership districts
Soil and water conservation districts
Sanitary districts
Sewer authorities
Lake improvement, conservation, and various other authorities
Watershed districts

flooding. Thus, although water is renewable, it is the most limiting factor in worldwide agricultural production (31) and for commercial manufacturing and domestic purposes in semiarid and arid regions. Already in some areas (Aurora, Colorado) local ordinances not only restrict water use for lawns, including turfgrass and landscape plants, but they also restrict the size of the lawn area (21).

Water uses

Both surface and underground water sources provide for humanity's many and varied uses. Water may be used "instream" or "offstream" or it may be subjected to "consumptive use."

Instream. Instream uses do not require removal or diversion from a creek, river, lake, or reservoir. All uses take place within or upon the surface. Examples of instream use include hydropower generation, navigation, and such various recreational and environmental activities as swimming, boating, fishing, maintaining water levels needed for aesthetic purposes and water quality, and avoiding intrusion by saline water, especially along coastal areas.

Offstream. Offstream uses include domestic, commercial, manufacturing (including steam and nuclear power generation), and agriculture categories. Water used for domestic and manufacturing purposes remains principally liquid and, when pollutants are removed, it may be used again, for the most part, for these or other purposes.

Consumptive use. Water used for irrigation in agriculture, including irrigation of turfgrass and landscaped areas, is incorporated into the plant or animal body, is lost as vapor to the atmosphere through evapotranspiration or percolates into "groundwater." This water, removed from the immediate water supply, is said to be "consumed."

Consumed water cannot be reused until it returns to earth as precipitation. A 1962 estimate stated that the water withdrawn from surface and subsurface sources for irrigation is equivalent to 3.4 percent of the amount that falls as annual precipitation on the conterminous 48 states; 60 percent of this amount is lost to the atmosphere (35). Also, an estimated 70 percent of precipitation falling annually on the United States is lost to the atmosphere from nonirrigated land areas—23 percent from crops and pasture, 16 percent from forest and browse areas, and 32 percent from noneconomic vegetation.

Volumes of water withdrawn and consumed in 1975 and estimates for 1985 and 2000 are shown in Table 1. These amounts are shown for the major functional use areas and are expressed in millions of gallons per day (mgd).

Total fresh water withdrawals in 1975 for all offstream uses listed in Table 1 were estimated at 338.5 billions of gallons per day (bgd) (28). By the year 2000, withdrawals are expected to decline 9 percent to 306.4 bgd. The most significant decline is expected in the manufacturing area where a 62 percent reduction is expected. Quantities withdrawn for irrigation and steam electric generation will decrease slightly from 1975 to 2000, although percentages for each will remain essentially unchanged at about 50 percent for irrigation and 25 percent for steam electric generation. All other offstream uses will increase in quantity and in percentages.

Water consumption is perhaps more critical than the total amount withdrawn because it is unavailable for downstream uses or recharge of underground reservoirs. Within the various use categories shown, only about one-third of the water withdrawn is consumed. The U.S. Water Resources Council (28) estimates that agriculture was responsible for 83 percent of the water consumed in 1975. This same source estimates that this percentage will decrease by the year 2000 because of expected increases in consumption for power generation (steam) and manufacturing. Consumptive use is expected to increase approximately 27 percent from 106.6 bgd in 1975 to 135.1 bgd in 2000.

The projected decline between 1975 and 2000 in the total amount of water withdrawn for agricultural purposes, including irrigation of turf and landscaped areas, will reflect, among other things: Depletion of irrigation water available from major aquifers like the Ogallala and those drawn upon to supply water to the San Joaquin Valley of central California (10, 34); more urbanization, which may result in the shift of available water from

TABLE 1 Estimated annual fresh water withdrawals and consumption for 1975, 1985, and 2000

	Million gallons per day					
	Total withdrawals			Total consumption		
Functional use	1975	1985	2000	1975	1985	2000
Domestic	23,256	26,303	30,318	6,268	7,073	8,074
Commercial	5,530	6,048	6,732	1,109	1,216	1,369
Manufacturing	51,222	23,687	19,669	6,059	8,903	14,699
Agriculture						
Irrigation	158,743	166,252	153,846	86,391	92,820	92,506
Livestock	1,912	2,233	2,551	1,912	2,233	2,551
Power generation	88,916	94,858	79,492	1,419	4,062	10,541
Minerals industry	7,055	8,832	11,328	2,196	2,777	3,609
Public lands and others	1,866	2,162	2,461	1,236	1,461	1,731
TOTAL FRESH WATER	338,500	330,375	306,397	106,590	120,545	135,080

SOURCE: U.S. Water Resources Council, 1978.

cropland irrigation to domestic use as could happen in the case of the El Paso, Texas area (18); curtailment of domestic water use for irrigation of lawns and other landscaped areas as is occurring in Aurora, Colorado (21, 22); and increasing costs of the energy needed either to pump water or to transport it to distant locations.

The total projected decline in fresh water withdrawal (9 percent) for all offstream uses assumes improved efficiency in water use and in recycling techniques; both are the result of anticipated improvement in technology, conservation, and management.

Hydrologic cycle

Water covers two-thirds of the earth's surface, yet less than 1 percent of this global supply is fresh water (3, 24). The remaining 99 percent is tied up in oceans (97 percent) and in polar ice caps (2 percent). This 99 percent is basically unavailable and unsuitable for humanity's primary uses—drinking; sanitation; agriculture; manufacturing; processing of food, feed, and fiber; and certain types of recreation (27).

The amount of water on earth is essentially nondestructible and fixed. The activities of human levels do not affect it. The water available today is the same water that was available to Adam and Eve; it is the same water in which John baptized Jesus. Or, as Joseph P. Rossillon, president of the Freshwater Foundation, says, "It titillates my fancy to think the water I used for shaving this morning well may be the water in which Cleopatra bathed." All of this water still exists, its billions of molecules undergoing continual dispersal. Certain types of volcanic activity may add small amounts of new water, but they are insignificant in the overall scheme.

Worldwide, some 80,000 cubic miles of water evaporate each year from oceans and another 15,000 cubic miles from lakes, streams, and land surfaces (11). Some 24,000 cubic miles of this falls on land surfaces. This is approximately equal to a depth of 475 feet over all of Texas (1) or enough to form a global lake 33 feet deep (27).

It is estimated that, on average, between 600 and 675 billion gallons of fresh water fall on the United States daily (13, 34). This represents approximately 3 percent of the world's total. Canada and the Soviet Union each have more than 20 percent (34).

Moisture evaporated from water or land sources and transpired by vegetation is lifted into the atmosphere as vapor (evapotranspiration). Sooner or later it condenses and falls to earth as rain, fog, sleet, snow, hail, or dew. This is a continuous, never-ending process of charging (evapotranspiration) and discharging (precipitation). It is a cycle—a solar-fired water cycling process known as the "hydrologic" or water cycle (fig. 1). It is this process that has caused the same water to be used and reused since time on earth began.

The amount of evaporation from all surfaces and the amount transpired by plants is approximately equal to the amount of water that falls as precipitation daily. Precipitation patterns, however, are widely variable and seldom, if ever, is the right amount of water, in any form, delivered to the place where it is needed, or not needed, as the case may be. Human activity, while often lowering water quality and frequently displacing water from natural channels, has little effect on the hydrologic cycle. In the future, through weather modification, humans may disrupt the hydrologic cycle or affect precipitation in a specific location. Laws covering this segment of the hydrologic cycle—atmospheric water—are, only now, being developed (26).

Water sources and supply

Precipitation that falls on land surfaces infiltrates the earth's surface to provide water for plants and other soilborne organisms, replenishes reservoirs of groundwater (aquifers), or runs off into lakes and streams, where it evaporates. Some returns to the oceans. Figure 2 shows the U.S. "water budget."

Some 40,000 bgd of water vapor passes over the contiguous 48 states. Approximately 10 percent (4,200 bgd) falls as precipitation. This amount is equivalent to an average annual rainfall

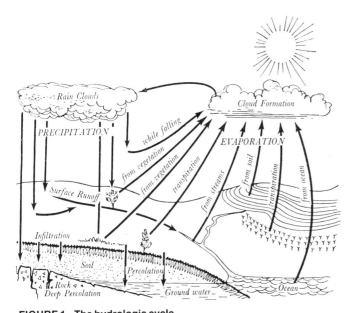

FIGURE 1. The hydrologic cycle
—Water, 1955 Yearbook of Agriculture

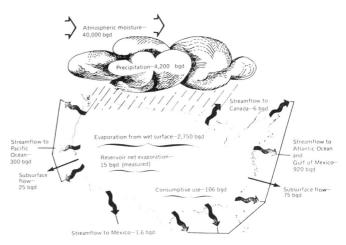

FIGURE 2. Water budget of the conterminous United States
—U.S. Water Resources Council

of 30 inches. Two-thirds of this amount (2,750 bgd) evaporates immediately or is transpired. The remainder (1,450 bgd) accumulates in surface or ground storage areas, is used consumptively, evaporates or runs off into the Atlantic and Pacific oceans, the Gulf of Mexico, or rivers and streams going into Canada or Mexico. Only 675 bgd of the potential for 1,450 bgd are considered available in 95 of each 100 years (28). Insufficient surface storage areas and the extreme variability of annual precipitation, which may result in floods or drought, can limit recapture or utilization of greater amounts of the potential volume of water.

Precipitation

Average annual precipitation patterns in the United States range from less than 10 inches in the West and Southwest to well over 100 inches in the Pacific Northwest with, as noted earlier, an average of 30 inches for the conterminous states (fig. 3). Two-thirds of the average 30 inches of annual rainfall, falls in the eastern half, the rest in the western. The approximate dividing line between East and West is along the 100th meridian. West of this line, annual evaporation exceeds precipitation. The division also falls along the line between the subhumid and semiarid zones. In subhumid and humid zones, precipitation exceeds evaporation. The line varies, but generally it runs through eastern North Dakota, western Minnesota, eastern South Dakota, east-central Nebraska, Kansas, Oklahoma, and Texas—terminating at approximately Corpus Christi.

It is apparent that the water-short western half must find additional sources of water if it is to grow, develop, and expand as it has in the past. Because of the unequal distribution of rainfall, 17 western states contain 85 percent of the land irrigated for agricultural purposes. Too, the Southwest (Texas, New Mexico, Arizona, and southern California) represents a very large portion of the Sun Belt, which is experiencing population growth. All must have water.

The geopolitical implications of the nation's unequal distribution of rainfall are enormous. Sooner or later the inequitable distribution and limited availability of usable water may well reverse recent population shifts. The legal and social implications of transporting water to correct already existing and future inequities are discussed in chapter 2, *Water: Whose is it and who gets it?*

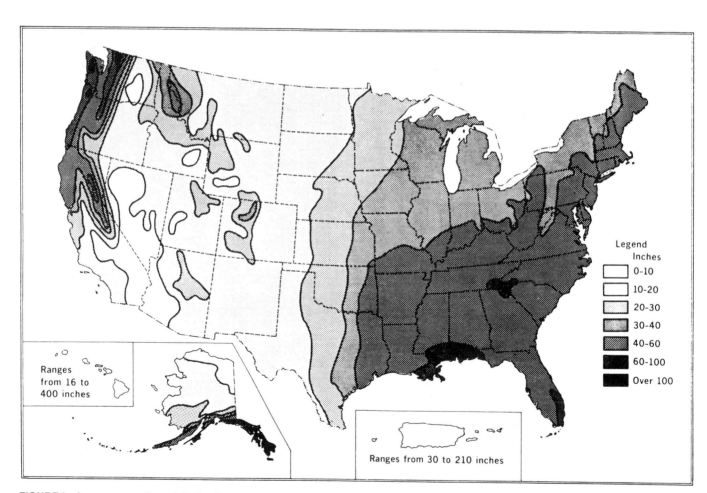

FIGURE 3. Average annual precipitation for the United States. Values for any given area are often misleading. Wide variability in rainfall occurs seasonally, annually, and, in some instances, in areas of proximity. An example of the latter occurs in Kuwai, Hawaii; Mount Waialeale averages more than 400 inches of rain annually; 25 miles to the southwest, on the leeward side of the island, rainfall is less than 20 inches (12).

—*U.S. Water Resources Council*

Runoff water

The average annual runoff for the United States is shown in figure 4. The amount of runoff parallels precipitation patterns, being lowest in the Southwest and intermountain valleys and highest in the Pacific Northwest and the East.

Runoff, theoretically, is water that could be added to the water supply. However, it is not possible to capture and store all runoff water. Additionally, some runoff is essential to maintain streamflow, lake and dam levels, and other natural storage areas. Nevertheless, a reduction in runoff from major storms would add to supplies of surface and underground waters. Capture and impoundment of runoff waters from surrounding watersheds represents a supplemental source of irrigation water for golf courses and other large turf areas when terrain is suitable and water laws permit.

Factors affecting runoff. The amount or volume of runoff on any given land area or from any given storm depends on several factors: intensity and length of the rainfall period, type of vegetative cover, steepness of slope, soil texture, and amount of water in the soil. In the case of snow, the depth of the snowpack, its rate of melt, and the depth of the frost line are factors.

In general, runoff increases as rainfall intensity and steepness of slope increase. Very dry silty-clay or clay-type soils frequently exhibit hydrophobic properties; on the other hand, soils, when saturated, or nearly so, are unable to accept much additional water. Thus, in both cases, runoff is greater. Storm or rainfall intensity is measured not only on the basis of volume (inches) per hour, but also on the amount of water the soil can absorb before reaching saturation (8).

Turf cover reduces runoff and protects soil surfaces from erosion by wind and water. Further, it improves the infiltration rate of clay and silty-clay soils and increases the water-holding capacity of sandy-type soils. Additionally, turf cover will filter atmospheric pollutants and particulate matter brought to earth by precipitation. There is a paucity of information documenting the specific and direct effects of turf cover as an erosion control agent. However, average annual soil and water losses from a number of soil series with variable slopes, rainfall, and crops were compared (8). Table 2 on page 26 is a selected compilation from the 20-plus examples presented. In addition, data were presented comparing soil and water losses as influenced by slope and by intensity of rainfall for several cropping situations.

Clearly, grass cover, compared with various cropping procedures, effectively minimizes soil and water losses. Only ma-

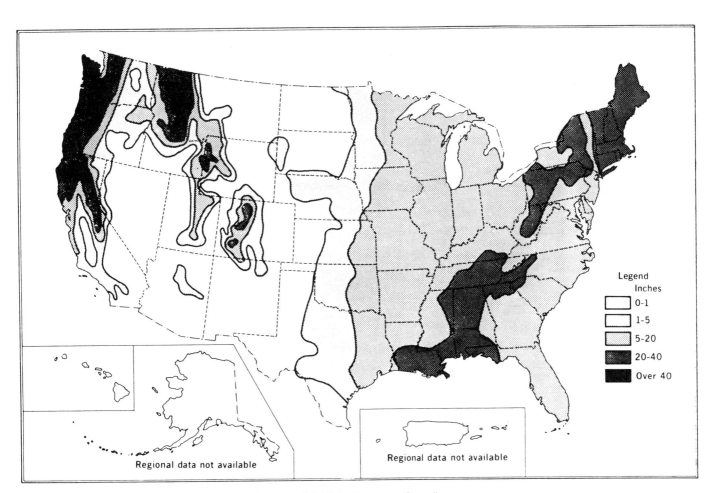

FIGURE 4. Average annual runoff for the United States —*U.S. Water Resources Council*

TABLE 2 Average annual soil and water losses from selected sites under different cropping systems

Soil series*	Shelby loam		Shelby silt loam		Clinton silt loam		Vernon fine sandy loam		Cecil sandy clay loam	
Percent slope	3.68		8		16		7.7		10.0	
Average precipitation (inches)	40.37		34.79		34.12		33.12		48.42	
Period of study (yrs inclusive)	1918–1931		1931–1935		1933–1935		1930–1935		1932–1936	
Crop or treatment	Tons of soil per acre	Percent precipitation	Tons of soil per acre	Percent precipitation	Tons of soil per acre	Percent precipitation	Tons of soil per acre	Percent precipitation	Tons of soil per acre	Percent precipitation
Fallow	41.6	30.7	112.79	31.20	151.94	20.44	20.30	27.45	64.68	29.13
Continuous	19.7	29.4	85.48	27.28	84.06	19.76	28.13	14.65	25.76	9.42
Rotation	2.7	13.8	11.35	15.23	25.31	12.07	5.5	11.58	10.84	8.81
Grass	0.3	12.0	0.29	9.30	0.03	2.82	0.03	1.23	0.014	0.29

SOURCE: Adapted from Bennett Soil Conservation

*Shelby loam: Continuous corn; corn, wheat, clover rotation; Kentucky bluegrass (*Poa pratensis*).
Shelby silt loam: Continuous corn; corn, wheat, clover rotation; continuous grass (species unknown).
Clinton silt loam: Continuous corn; corn, barley, clover rotation; and Kentucky bluegrass (*Poa pratensis* L.).
Vernon fine sandy loam: Continuous cotton; cotton, wheat, sweet clover rotation; clipped bermudagrass (*Cynodon dactylon*).
Cecil sandy clay loam: Continuous cotton; wheat, corn, cotton, lespedeza rotation; mixed grasses (species unknown).

ture forests are as effective, or slightly more so, in controlling erosion (8). Data to illustrate the comparative effectiveness of turfgrass in erosion control are unavailable; however, it is reasonable to assume that results would be similar to those for various grasses. One would expect, therefore, that dense, weed-free, properly maintained lawns and other turfgrass facilities would reduce soil and water losses. Also, properly applied fertilizers would not be expected to pollute adjacent or nearby lakes and streams. In fact, good, properly maintained turfgrass lawns, with congruous size, soil, and slope conditions will protect lakes and streams against nutrient enrichment from nearby croplands.

Surface water

Surface water is water found in lakes, rivers, streams, canals, swamps, artificial ponds, and impoundments. Most surface water areas depend on runoff for replenishment. This varies widely from region to region (fig. 4), from season to season, and even from day to day. Within any given year for any given stream, maximum flow may be as much as 500 times greater than minimum flow (28). The range is generally less than this; nevertheless, for those areas dependent upon a high proportion of available average runoff, a drought, particularly a protracted one, may be very serious. Variation is less on an average daily or annual basis in the humid East. Even so, in humid areas, severe drought may draw down surface water supplies (as well as underground ones) to alarming levels. Less than a 30-day supply for major metropolitan areas has been experienced. These situations result in restrictive watering of all turfgrass areas including golf courses with their own water sources (5, 19, 29). To date, rains have come in time to replenish reservoirs, but as use expands, emergency situations may well occur from the exhaustion of water supplies.

In humid and subhumid areas watering of home lawns often is restricted because municipal distributive systems have not kept up with the rapid expansion of suburban areas. For the same reasons in semiarid and arid regions watering of turfgrass areas may be restricted. In addition, in these areas, restrictions are often the result of either inadequate supplies or an inability to obtain new surface-water rights at reasonable costs. When water rights are available or when runoff water can be impounded on site, the turfgrass facility's water supply will increase.

Small reservoirs (less than 50 acre-feet in size) provide only 2 percent of the nation's total storage capacity (Table 3); yet, they are a significant source of water for golf courses and park areas. Water harvesting and storage in small ponds and reservoirs are increasingly major elements in golf course design (2, 5). Reservoir storage capacity for the United States in 1978 is shown in Table 3.

Surface waters of all types are common direct sources of water for golf courses and other larger turfgrass areas. Frequently, small streams and major drainage channels may be dammed, excavated, or both, and the impounded water used to

TABLE 3 Storage capacity of reservoirs in the United States

Size in acre-feet	Number	Total usable storage capacity (million acre-feet)	Total storage capacity all reservoirs (percent)
Greater than 2 million	31	191	41
5,001 to 2 million	1,600	159	37
50 to 5,000	47,500	90	20
Less than 50	1,800,000	10	2
TOTAL	1,849,131	450	100

SOURCE: U.S. Water Resources Council, 1978

Small lakes add to the aesthetic appeal of a turf area. In the case of golf courses, they add to the playing strategy of the hole.
—Photo courtesy The Toro Company

irrigate the golf course (7, 17, 25). Increasingly, many real estate developments utilize low areas or floodplains to locate their golf courses, parks, and recreational areas (17). On floodplains the major problem may be that of proper channeling or diking to avoid flooding. On many sites, however, the major problem is one of drainage to accommodate excess precipitation and to lower water tables. Drainage water, when properly channeled, diverted, or directed to on-site impoundment areas, may be utilized to supplement future water needs and recreational uses for the turfgrass facility.

When properly planned, drainage, runoff, and surface waters, including strategically located streams and impoundments, may be incorporated as an integral part of the park, playground, or golf course design (6, 7, 17, 25). As such, they enhance the aesthetic appeal and, in the case of the golf course, represent a physical feature that the architect may employ to improve the play of the hole or holes located adjacent to the water. Additionally, ponds and reservoirs located on turf facilities provide beneficial instream recreational possibilities, such as fishing and boating.

Effluent water, although not technically "surface water," is a potential or alternative source of supplemental irrigation water when impounded after treatment. Because of its plant nutrient content, it is a particularly valuable source of irrigation water for sod farms and for golf courses. Advantages and disadvantages are associated with use of effluent (30). Effluent or wastewater is used to irrigate several golf courses; among them are the Eisenhower Course at the Air Force Academy; Innisbrook at Tarpon Springs, Florida; Randolph Park at Tucson, Arizona; and some military courses.

Significant to this discussion is the fact that wastewaters are a source of water for turfgrass facilities. In fact, they are ideally suited because of the filtering action of grass and soil. In some semiarid and arid areas, restrictions have been placed on their removal from stream flows. If they are removed, the user must compensate municipalities.

Golf course water requirements. Golf courses may require a water source capable of supplying as much as 1.5 to 3.5 million gallons of water per week (17). Approximate capacity needed for specific locations may be calculated from potential evapotranspiration, modified by long-term rainfall data (4). Water-use rates vary widely. Variations from 0.1 inch per day for foggy coastal climates to 0.45 inch per day for dry desert areas have been reported (7). This equates to 0.7 and 3.1 inches per week or 19,000 and 84,000 gallons per week, respectively. (See conversion table for water-measurement terms.)

Failure to utilize fully the various sources of surface and wastewaters to irrigate turf facilities, especially golf courses, will hasten the day when legal curtailment of water, ranging from rationing to complete prohibition, will become a reality.

Groundwater

Most underground water is precipitation that has seeped down from the surface. Water that seeps into aquifers (reservoirs of

Lakes and reservoirs add aesthetic appeal and provide a source of water for irrigation of golf courses.
—Photo courtesy The Toro Company

Impoundment of runoff water and damming of small streams provide a source of water for irrigation of large turfgrass areas. Grassy edges control erosion.
—Photo courtesy The Toro Company

CONVERSION TABLE FOR WATER-MEASUREMENT TERMS		
VOLUME		
1 acre-inch	=	43,560 cubic inches
	=	27,154 gallons
	=	amount of water needed to cover 1 acre to depth of 1 inch
1 acre-foot	=	43,560 cubic feet
	=	325,851 gallons
	=	amount of water needed to cover 1 acre to depth of 1 foot
1 gallon	=	0.134 cubic feet
7½ gallons	=	231 cubic inches
	=	1 cubic foot
1 million gallons	=	3.07 acre-feet
1 billion gallons	=	1,000 million gallons
FLOW—VOLUME		
1 million gallons per day (mgd)	=	694.4 gallons per minute
	=	1.5 cubic feet per second
	=	1,120 acre-feet per year
1 billion gallons per day (bgd)	=	1.12 million acre-feet per year
1 cubic foot per second	=	1.98 acre-feet per day
WEIGHT		
1 gallon of water at 62°F	=	8.34 pounds
1 cubic foot of water at 39.1°F	=	62.43 pounds
1 pound of water at 62°F	=	0.1199 gallons

underground water) is what's left after evapotranspiration or after it has been held in the vadose water zone. The depth and extensiveness of the aquifer is governed by climate (rainfall and EVT) and by the nature and character of the rocks or unconsolidated clays, silts, sands, and gravels in which it collects. Water moves into (infiltrates) surface soils relatively easily. Cracks and fissures in surface rocks, texture and structure of surface soil, and the degree of surface soil compaction, slope, and vegetative cover are major factors controlling the rate of infiltration.

Once into the soil, water moves through (percolates) the vadose water zone in accordance with, and consistent with, the permeability of the zone, including cracks, fissures, and channels, of the material through which it is moving. When it reaches bedrock or when, because of depth, the resulting pressure causes the parent material, consolidated rock, or soil to become too dense and thereby impervious, the water stops and, for all practical purposes, ceases to move more deeply. This usually occurs at a depth of about 2,000 to 3,000 feet. Most aquifers or reservoirs of underground water are located within 2,500 feet of the surface (28).

When water reaches its maximum percolation depth, it accumulates and saturates the water-bearing rocks, soil, clay, sand, or gravel, as the case may be. If this saturated zone is permeable enough to yield water in useful quantities, it is called an aquifer. Over time, the saturation zone moves upward toward the surface. The top level of the saturated zone is called the water table, and it fluctuates in accordance with rainfall and the rate and amount of water removed from the aquifer, especially through wells. The water table is rarely static, level, or equally distant from the surface for any significantly sized area.

When surface land areas reach down below the water table and into the zone of groundwater saturation, water will flow or ooze out through an opening. If the opening is small, the outflow is called a spring, which may form a branch or brook. If the surface is a deep valley extending well below the water table, a lake is formed, and if the surface extends only a short distance into the zone of saturation or is only near the water table, swamps and bogs are formed. About 30 percent of streamflow in the United States, in an average year, is supplied by springs, lakes, and swamps (28). Seepage from these sources and, in turn, from canals is a source of groundwater recharge. This interconnection may be beneficial, in the case of maintenance of streamflow, or detrimental, in the case of polluted surface waters that may move into groundwater. In turn, of course, polluted groundwaters may invade surface waters. Groundwater pollution by surface waters is particularly serious in karstic and other limestone regions because of their low-filtering capability and the fact that water may move laterally for long distances as it seeks to move downward.

The major United States aquifers are shown in figure 5.

Aquifers may be extensive (the Ogallala) or very local (the Jordan in Minneapolis, Minnesota), near the surface or hundreds of feet below it. Some may be closed (landlocked); others flow slowly, very slowly, but are usually parallel to the surface gradient. They may be very thin, sometimes just a lens of fresh water overlying salt water, or hundreds of feet thick. Some contain brackish water, and some have become saline as a result of saltwater intrusion.

Aquifers are recharged primarily by precipitation—rain and snowmelt—and, as indicated above, in part by seepage. When recharge is less than the rate of withdrawal, the aquifer is being "mined"—overdrawn. Such is the case, at present, with several major U.S. aquifers. Table 4 shows the percentage of overdraft in each water resource region and the number of subregions with overdraft—60 of the 106 (figs. 6 and 7, pages 30 and 31, respectively). This degree of water mining has serious implications for those aquifers located in regions of low rainfall because they may not be recharged.

In the conterminous states, about 86 percent of the water

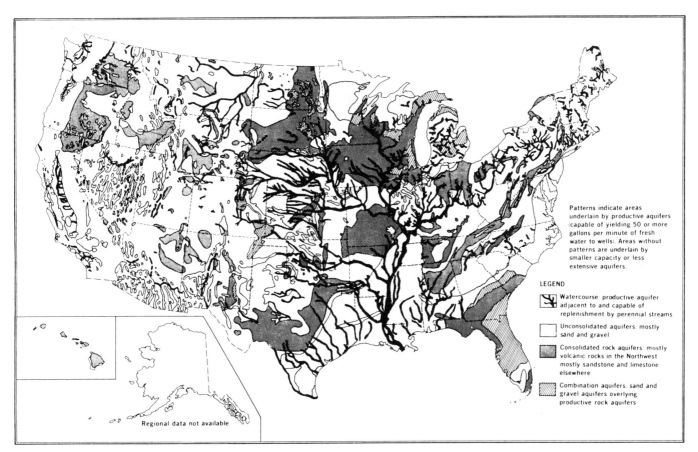

FIGURE 5. Major aquifers in the United States —*U.S. Water Resources Council*

TABLE 4 Approximate groundwater withdrawals and percentage of overdraft, 1975

Water resources region and number	Total withdrawal (mgd)	Overdraft (Reg's.) Total (mgd)	(%)	Subregions No. in region	No. with overdraft	Range in overdraft (%)
New England (1)	635	0	0	6	0	—
Mid-Atlantic (2)	2,661	32	1.2	6	3	1–9
South Atlantic/Gulf(3)	5,449	339	6.2	9	8	2–13
Great Lakes (4)	1,215	27	2.2	8	1	30
Ohio (5)	1,843	0	0	7	0	—
Tennessee (6)	271	0	0	2	0	—
Upper Mississippi (7)	2,366	0	0	5	0	—
Lower Mississippi (8)	4,838	412	8.5	3	3	7–13
Souris-Red-Rainy (9)	86	0	0	1	0	—
Missouri (10)	10,407	2,557	24.6	11	10	4–36
Arkansas-Red-White (11)	8,846	5,457	61.7	7	7	2–76
Texas-Gulf (12)	7,222	5,578	77.2	5	5	24–95
Rio Grande (13)	2,335	657	28.1	5	4	22–43
Upper Colorado (14)	126	0	0	3	0	—
Lower Colorado (15)	5,008	2,415	48.2	3	3	7–53
Great Basin (16)	1,424	591	41.5	4	4	7–75
Pacific N.W. (17)	7,348	627	8.5	7	6	4–45
California (18)	19,160	2,197	11.5	7	5	7–31
Regions 1 – 18/TOTAL	81,240	20,889	25.7	99	59	1–95
Alaska (19)	44	0	0	1	0	—
Hawaii (20)	790	0	0	4	0	—
Caribbean (21)	254	13	5.1	2	1	5
Regions 1 – 21/TOTAL	82,328	20,902	25.4	106	60	1–95

SOURCE: U.S. Water Resources Council, 1978

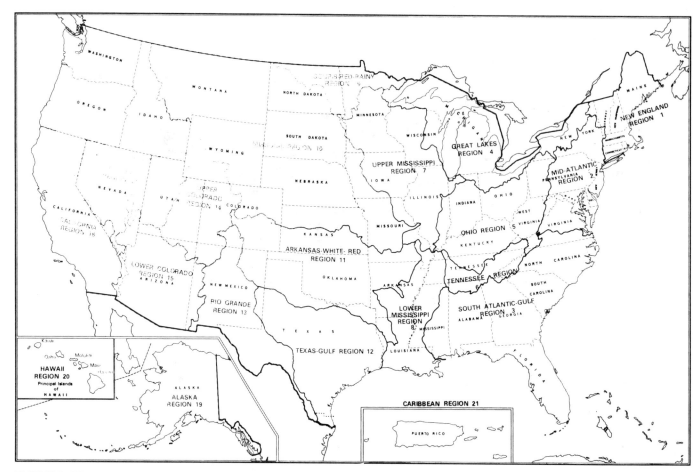

FIGURE 6. Water resources regions in the United States —*U.S. Water Resources Council*

resources is located in aquifers and is the source of approximately 20 percent of the fresh water used in the United States. This use has been increasing at an annual rate of 3.8 percent (28).

The amount of groundwater resources in the conterminous 48 states is much greater than the available water in lakes and streams. Estimates vary from 33,000 trillion to 59,000 trillion gallons (28). Compare this with the 1,290 trillion gallons in Lake Michigan or with the 33,000 trillion gallons discharged by the Mississippi River into the Gulf of Mexico during the last 200 years. Because of costs and environmental concerns, all of the groundwater should not be considered as available for our use.

Overdraft and its consequences

Use of fresh groundwater has been increasing annually at a rate of 3.8 percent on average (28). About 68 percent of the total withdrawn each year is used for irrigation. This represents approximately 35 percent of the water used for this purpose (28). Significant is the fact that in some areas groundwater is being withdrawn at a greater rate than it is being replenished. Three examples will illustrate problems arising from overpumping of groundwater. They are California's San Joaquin Valley, the El Paso, Texas area, and the Ogallala aquifer.

San Joaquin Valley. In 1975, 48 percent of the total fresh water used in California came from wells (28). This amount equaled 23 percent of the total amount of groundwater withdrawn nationwide and represents an amount greater than that extracted by all eastern regions combined.

Overpumping of one area in California, the San Joaquin Valley aquifers, is estimated at 1.5 million acre-feet of water per year (27). As a result, the land has subsided several feet during the last quarter century. More importantly, soil and water quality problems are becoming serious, as a result of the salts the water collects as it percolates through the soil and those that remain when salty capillary water moves to the surface and evaporates.

Land subsidence has occurred in other areas; for example, in a suburb of Houston, Baytown, Texas, overpumping caused not only land to subside but resulted in an invasion of tidewater from Galveston Bay (10).

El Paso area. A different problem faces El Paso. Its water comes from relatively small aquifers (bolsons) located in and near the Rio Grande floodplain. The city obtains water from some of these underground sources and from the Rio Grande (fig. 8, page 32 (18)).

The population of Texas in 1980 was 14,230,000, of which 11,330,000 (or 80 percent) was urban (18). Annual water use for

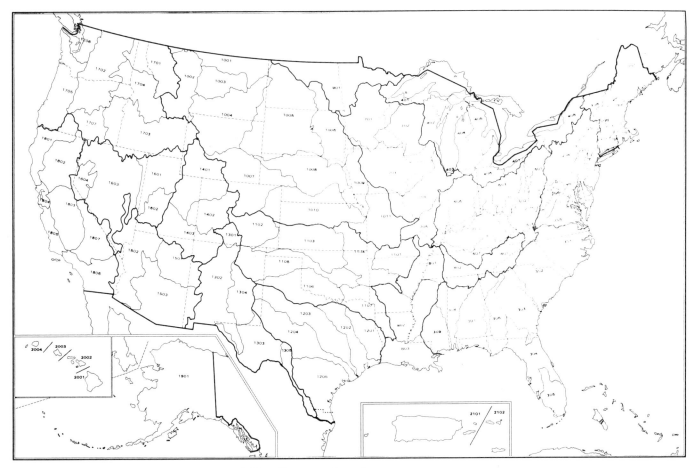

FIGURE 7. Water resources subregions —*U.S. Water Resources Council*

the state in 1980 was reportedly 19 million acre-feet, up from 2 million acre-feet in 1930.

Population trends in El Paso County are shown in Table 5. Rapid growth in population has placed heavy demands on the city's water supply. Reported sources of water for El Paso's domestic use are shown in Table 6.

Future domestic water needs portend problems for many communities. Table 7 illustrates costs and problems associated with potential water sources for El Paso (18).

Clearly, removal of the legal restrictions presently limiting use of the Mesilla bolson and additional water from the Rio Grande would temporarily relieve El Paso's pending problems. This, however, may not be the best solution for El Paso's neighbors, and, in this case, is involved the question of state ownership of water since this aquifer extends into New Mexico. Recent litigation has established, for the first time, that indeed El Paso, Texas may drill additional wells into that portion of the Mesilla

TABLE 5 Population trends, El Paso County (18)

Year	Population	Percent living in city
1940	96,800	74
1950	130,500	67
1960	277,000	88
1970	322,000	90
1980	480,000	90
1990 (est.)	570,000	?

TABLE 6 Domestic water sources (acre-feet per year), El Paso (18)

Year	Rio Grande	Underground	Total
1974	14,200	97,700	111,900
1980	13,500	123,500	137,000

TABLE 7 Water sources and costs, El Paso (18)

Source	Cost per 1,000 gal	Problems
Brackish HUECO Bolson	.30	Salty
Expanded use HUECO	.30	Exhaust aquifer
More Rio Grande	.70	Legal restrictions
Mesilla Bolson	.90	Legal restrictions
Recycle sewage	$1.20	Energy, pumping
Fresh water within 150 miles	$1.50	Energy, health effects
Desalinization	$2.40	Energy, brine disposal; poor well yields

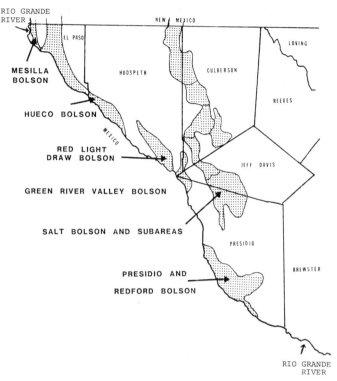

FIGURE 8. Water sources for El Paso, Texas (18)

from the ancient Rocky Mountains during the Pliocene and early Pleistocene periods. During those periods, also, the detritus became saturated and eventually water was trapped. The Ogallala overlies primarily an impermeable shale, some limestone and some sandstone, and is capped by consolidated sandstone, some limestone as well as chalky marl and shale. These caprocks generally resist erosion and are only slowly permeable—permeable enough, however, to accept the estimated ½ inch to 3 inches of recharge water remaining from the average annual rainfall, which ranges from 12 to 22 inches, with an average over the entire formation of about 18.2 inches (12). Before large-scale overpumping, there was enough recharge to maintain the Ogallala in a balanced state.

Depletion

In the High Plains section of the Texas Panhandle, before 1936, the 300 wells sunk into the Ogallala irrigated some 35,000 acres. By 1941 there were 2,560 operating wells and in 1948 there were

bolson underlying the state of New Mexico to obtain water for domestic use. (See chapter 2, *Water: Whose is it and who gets it?*)

Another alternative may be to purchase "water rights" from surrounding landowners. However, lands along the Rio Grande are highly productive, and water from nonfarm lands may be too distant to be economically practical at this time. Furthermore, it may be economically impractical for domestic use, more specifically for human consumption. The upper limits of cost for this purpose may be illustrated by the price some pay today for "spring" and distilled water—90 cents to $1 per gallon. This, by no means, is the upper limit of water's value!

The Ogallala. The Ogallala aquifer, named for a Native American tribe who once roamed parts of the Great Plains of North America, lies beneath parts of eight states. It starts in South Dakota and flows south by southwest to the High Plains of Texas (fig. 9).

The borders are very irregular and the water-bearing strata of the Ogallala vary in thickness from under 100 feet to over 1,000 feet with an average thickness of 200 to 300 feet. Depth of the water table also varies from a few feet—less than 50—to as much as 150 to 250 feet or more. The Ogallala formation, estimated to contain in excess of 2 billion acre-feet of water (2, 16), is perhaps the largest underground reservoir of fresh water in the world.

The Ogallala's water-bearing strata consist of unconsolidated detritus, principally gravel, sand, silt, and clay eroded

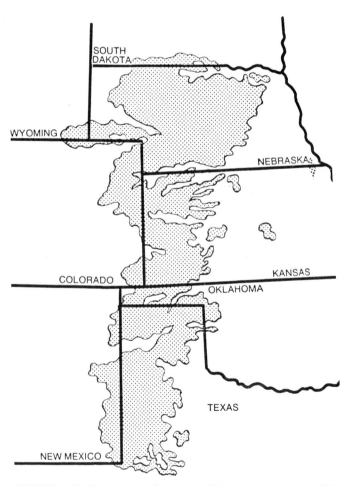

FIGURE 9. The Ogallala aquifer cuts a path through eight states. This vast aquifer is the sole source of water for most of the High Plains region.

—*The Denver Post*, 1979

This center pivot system, located in the Great Plains, uses water from the Ogallala aquifer. Note condition of nonirrigated area in background.
—Photo courtesy The Toro Company

Grain is being grown by center pivot systems in Ogallala area.
—Photo courtesy The Toro Company

8,500 wells irrigating approximately a million acres (11). Today there are over 70,000 wells irrigating in excess of 6 million acres (34). These fields produce 25 percent of the nation's cotton (34) and millions of pounds of grain, which are used as feed for 40 percent of the nation's fed beef (13).

The overdraft in the Texas-Oklahoma High Plains alone is estimated at 14 million acre-feet annually (more than 12,500 mgd). This is an amount of water equal to the annual flow of the Colorado River (28). Obviously, the Ogallala is being mined and exhausted of water that fell thousands of years ago!

Similar expansion of irrigated farmlands has occurred in the other seven states overlying the Ogallala, and similar overdrafts are resulting. Despite the immense importance of these areas to feed the nation, their decline is inevitable. Already, irrigated acreage has begun to decrease and some estimates anticipate a reduction to 3.2 million acres from 1980's 6-plus million (34). By the turn of the century, at present rates of use, southwest Nebraska, eastern Colorado, and western Kansas will have exhausted the recoverable underground water from the Ogallala (34).

Alternatives

What can be done to prolong the life of the Ogallala or of irrigated agriculture in America's vast heartland? Schemes abound! Among them: Diversion of water from the Missouri River and some of its tributaries to recharge the aquifer; canals or pipelines from the Arkansas and Mississippi rivers to bring water into the High Plains of Texas or, perhaps, divert part of the flow from the water proposed to be brought south from the Yukon River to disperse through the Great Plains.

Other, more plausible proposals (and far less complicated legally) suggest injection of water from playa lakes through wells into the aquifer. This would recharge and salvage large quantities of water normally lost to evaporation. A second proposal is to inject air through a well below the cap rock and thereby, because of air pressure, cause capillary water to move downward to the water table and replenish the aquifer (16). This scheme has been field-tested and appears economically feasible. One field study, the Idalou project, showed a net increase within 2,800 feet of the injection well (total area of 565 acres) of 225 acre-feet of water. Water levels measured in wells near the site indicated 315 acre-feet of additional water 100 days later and 406 acre-feet additional after 160 days. If this 406 acre-feet were recovered capillary water, cost of its recovery would be about $50 per acre-foot (16). Five other recovery techniques were tried; only air injection proved economically feasible.

The air injection technique obviously further depletes the aquifer and over time could possibly cause deformation of the water-bearing strata and preclude future natural recharge. The program may need legal justification; recovery of this natural exhaustible resource is, however, no different from recovery of minerals through mining operations or recovery of petroleum by use of similar techniques.

Improved water management procedures like well spacing, a long-time practice in the High Plains of Texas (14), and techniques of watering, such as surge irrigation for furrow irrigators (15), along with low-pressure sprinkler heads for center-pivot systems and other conservation practices, will help to conserve this immense resource.

Groundwater implications for turfgrass

Wells, common sources of water for golf courses and other turfgrass facilities, are in most cases discharged into a lake or pond; water is then pumped into the irrigation system for the golf course. Sometimes, wells may be connected directly to irrigation systems, although this is not generally the preferred procedure. The depth to the water table, the thickness of the aquifer, the volume of flow, and the proximity to salt or brackish water must all be taken into account. Also, increasingly, permits to drill wells are mandatory and inevitably will be so for all wells in all areas.

Wells provide a large part of the water for municipal systems, which in turn, provide water for most home lawns as well as other turf facilities. Restrictive uses for this source are similar to those for surface waters.

Concerted action by the green industry is needed to ensure fair allocation of water for turfgrass. Steps similar to those taken by turf industry members (29) to advise the South Florida Water Management District authorities on the role of turfgrass in the community, especially golf courses, need to become routine for all turfgrass associations. Reports were prepared showing the importance of the district's 450 golf courses to the economy and to the environment. Also, they indicated acreage for the five basic areas of the course—greens, tees, fairways, roughs, and common or nonplaying areas—the inches of water needed per week, and priorities. This type of approach, along with careful monitoring of the water source for the turf facility, will help to ensure that turfgrass areas will receive a fair allocation of tomorrow's water—never at the expense of human needs but certainly of high priority!

Water resources, regions, and subregions

The U.S. Water Resources Council (28) has divided the nation into 21 major water resource regions and, in turn, 106 subregions. Water resource regions are shown in figure 6 and subregions are shown in figure 7.

These classifications were established to assess the nation's water resources. Regions include the drainage area of a major river, as the Missouri (Region 10), or of several rivers, as the South Atlantic (Region 3). Subregions are smaller drainage areas used as basic data collection units and to point up basin-wide problems. In the aggregate, they reflect the wide contrast in both regional and national water sources and uses.

Subregion information is useful to turfgrass managers who wish to monitor their water uses and to begin to build data bases for future water needs for their facilities. The information obtainable is helpful in assessing the suitability of future facilities for various purposes.

Each region is headed by a commissioner responsible for collecting public information.

Summary

The United States has an abundant, currently adequate, and generally dependable supply of fresh water from surface and groundwater sources. On the average, in excess of 600 billion gallons per day of fresh water are available for beneficial uses.

Major problems are associated with the unequal distribution of precipitation, with its attendant problems of location and availability to population centers on a nationwide basis. The western United States receives only one-third of the nation's average annual rainfall, yet uses enormous quantities of water, especially for irrigation. The West has abundant energy reserves in the form of shale oil and coal. Vast quantities of water are needed to process these and, perhaps, transport them to the energy-poor East and Midwest. Specific examples of the problem of location, other than those discussed, are given by Wiegner (34). She points out that in Louisville, Kentucky, the groundwater table rose 32 feet between 1974 and 1979; at the same time in parts of Arizona the water table dropped 400 to 450 feet.

This inequitable distribution and location of available water supplies at points far distant from where they are needed or may be used beneficially for large numbers of people will inevitably give rise to complex social, economic, and legal problems. Certainly, water will cost much more in the future. And, in all likelihood, it will be transported or otherwise relocated far distances to support human and domestic needs.

Conservation and recycling of water, along with improved management and technology, will contribute to more efficient use. Research that leads to a better understanding of our water resources and how to manage them, coupled with development and utilization of drought and salinity-tolerant plants, will contribute further to wise use of our water.

Finally, programs and facilities that will lead to widespread use of both potable and nonpotable waters will conserve water and reduce its costs. Turfgrass interests, especially golf courses, are turning increasingly to irrigating with wastewaters. Others must follow suit. Development of dual-plumbing systems to accommodate both classes of water use for turf facilities, including home lawns, is inevitable. Most golf courses already use such systems.

Economic, social, and environmental constraints arising from water demands need not be incompatible with each other. These interrelationships are complex and their resolution difficult, however. And, as the pressures resulting from an increasing population and expanding technological and agricultural needs become more intense, the need to conserve and wisely use the nation's bountiful supplies of fresh water cries out for reasonable resolution.

The nation's turfgrass facilities are a national treasure and must not be overlooked in the development and execution of water policy. They contribute to the economy and recreational value of our communities and they are aesthetically appealing and functionally beneficial.

Literature cited

1. ACKERMAN, W. C., E. A. COLMAN, and HAROLD OGROSKY
    1955. Where We Get Our Water: From Ocean to Sky to Land. Yearbook of Agriculture. USDA Washington, D.C. pp. 41–51.
2. ADLER, V.
    1981. The Browning of America. Newsweek, February 23, pp. 26–37.
3. ALEXANDER, CALVIN
    1981. Water Is Where You Find It. Proceedings, 53rd International Turfgrass Conference and Show. New Orleans, La. GCSAA, Lawrence, Kans. pp. 1–2.
4. ANONYMOUS
    1966. Rainfall-Evapotranspiration Data—United States and Canada. The Toro Company, Minneapolis, Minn. 63 pp.
5. BAILEY, DAVID M.
    1982. Regional Response to the Water Challenge. Proceedings, 63rd International Turfgrass Conference and Show. New Orleans, La. GCSAA, Lawrence, Kans. pp. 3–4.
6. BEARD, JAMES B.
    1973. Turfgrass Science and Culture. Prentice Hall, Englewood Cliffs, N.J. 658 pp.

7. BEARD, JAMES B.
   1982. Turf Management for Golf Courses. Burgess Publishing Co., Minneapolis, Minn. 642 pp.
8. BENNETT, HUGH HAMMOND
   1939. Chapter 5, pp. 125–168. Rates of Erosion and Runoff Soil Conservation. McGraw-Hill, New York, N.Y. 993 pp.
9. BLAKE, GEORGE R.
   1980. Five-Year Water Resources Research and Development Plan. Goals and objectives. Water Research Center, University of Minnesota, St. Paul, Minn. 82 pp.
10. CANBY, THOMAS Y.
    1980. Water: Our Most Precious Resource. National Geographic (August). Vol. 158, No. 2. pp. 144–179.
11. CARHART, ARTHUR H.
    1951. Water or Your Life. J. B. Lippincott Co., New York, N.Y. 312 pp.
12. CLIMATE AND MAN
    1941. Yearbook of Agriculture. USDA, Washington, D.C. 1248 pp.
13. COOK, GAY
    1980. The Thirsty High Plains. Journal of Freshwater, Vol. 4. Special Issue: A Decade for Decisions. The Freshwater Foundation. Navarre, Minn. pp. 513–517.
14. CROSS SECTION, THE
    1982. Patricia Bruno, ed. Vol. 28, No. 11 (Nov). High Plains Underground Water Conservation District No. 1. Lubbock, Tex. 4 pp.
15. ———
    1982. A. Wayne Wyatt, mgr. Vol. 28, No. 12 (Dec). High Plains Underground Water Conservation District No. 1 Lubbock, Tex. 4 pp.
16. ———
    1983. A. Wayne Wyatt, mgr. Vol. 29, No. 3 (March). High Plains Underground Water Conservation District No. 1. Lubbock, Tex. 4 pp.
17. JONES, REES L., and GUY L. RANDO
    1974. Golf Course Development. The Urban Land Institute Tech. Bull. 70. Washington, D.C. 103 pp.
18. MALSTROM, HOWARD L.
    1982. An Overview of the Rio Grande Basin. Speech. Proceedings of the Southwest Turfgrass Conference. New Mexico State University, Las Cruces, N.Mex. 61 pp.
19. MARTIN, JOHN G.
    1982. The Water Crises in New Jersey—Dealing with the Bureaucracy. Proceedings, 53rd International Turfgrass Conference and Show. New Orleans, La. GCSAA, Lawrence, Kans. 105 pp.
20. OLSENIUS, CHRISTINE
    1980. Water in the U.S.A. Troubled Waters Ahead. Journal of Freshwater. Vol. 4. Special Issue: A Decade for Decisions. The Freshwater Foundation. Navarre, Minn. pp. S7–S9.
21. OLSON, STEVE
    1983. The Aurora Turfgrass Water Conservation Program. 29th Annual Rocky Mountain Turfgrass Conference. Colorado State University, Ft. Collins, Colo.
22. RONDON, JOANNE
    1980. Landscaping for Water Conservation in a Semi-Arid Environment. Department of Utilities, City of Aurora, Colo. 95 pp.
23. ROSSILLON, JOSEPH P.
    1981. Water in the 80's: The Decade for Change. Futurics. Vol. 5, No. 1. Special Issue: Water, Human Values and the 80's. A Consortium Effort to Keep Our Waters Usable. Pergamon Press, New York, N.Y. pp. 25–30.
24. ———
    1982. Water Problems in the 80's—Will Golf be Affected? Proceedings, 53rd International Turfgrass Conference and Show, New Orleans, La. GCSAA, Lawrence, Kans. pp. 67–69.
25. SEAY, EDWIN B.
    1982. New Design in Today's Limited Water Environment. Proceedings, 53rd International Turfgrass Conference and Show. New Orleans, La. GCSAA, Lawrence, Kans. pp. 73–77.
26. THOMAS WILLIAM A.
    1981. Allocation of Water as a Scarce Resource. Futurics. Vol. 5, No. 1. Special Issue: Water, Human Values and the 80's. A Consortium Effort to Keep Our Waters Usable. Pergamon Press, New York, N.Y. pp. 31–37.
27. THOMPSON, KAREN
    1982. Food for the Table: Will Water Set the Price? Journal of Freshwater. Vol. 6. Special Report. The Freshwater Foundation. Navarre, Minn. 32 pp.
28. U.S. WATER RESOURCES COUNCIL
    1978. The Nation's Water Resources, 1975–2000. Second National Water Assessment. Vol. 1: Summary. Superintendents of Documents, Washington, D.C. 86 pp.
29. WAGNER, BILL
    1981. Water Problems. The Florida Green. Winter. Florida Golf Course Superintendents Association, Lake Worth, Fla. p. 29.
30. WASTEWATER CONFERENCE PROCEEDINGS
    1978. James R. Watson, ed. Sponsored by: USGA, Green Section, Golf House, Far Hills, N.J.; Golf Course Architects Society, Chicago, Ill.; National Golf Foundation, North Palm Beach, Fla.; Golf Course Superintendents Association, Lawrence, Kans. 192 pp.
31. WATER USE IN AGRICULTURE: NOW AND IN THE FUTURE
    1982. Council for Science and Technology. Report No. 95. Ames, Iowa. 28 pp.
32. WATSON, JAMES R.
    1980. Future of the Turf Industry in the 80's. Proceedings of the New York Turfgrass Conference. Cornell University, Ithaca, N.Y.
33. ———
    1977. Drought Strategy for the Green Industry. Proceedings of the Texas Turfgrass Conference. Texas A & M University. College Station, Tex.
34. WEIGNER, KATHLEEN K.
    1979. The Water Crisis: It's Almost Here. Forbes, August 20. New York, N.Y. pp. 56–63.
35. WOLMAN, A.
    1962. Water Resources. National Academy of Sciences. National Research Council Publication 1000–B. Washington, D.C. 35 pp.

Additional references

BEARD, JAMES B.
  1981. Water-Conserving Turf Maintenance Practices. (First of a six-part series on water conservation.) Grounds Maintenance. March. Overland Park, Kans. pp. 16–26.
BROWN, JOHN W.
  1981. Project Leader. Evaluation of Agricultural Irrigation Projects Using Reclaimed Water. Boyle Engineering for the Office of Water Recycling, Water Resources Board. Sacramento, Calif. 225 pp.
ENERGY, WATER, ENVIRONMENT
  1977. Interview with Cecil D. Andrus, Secretary of the Interior. U.S. News and World Report, Inc., June. New York, N.Y. pp. 62–64.
GRAHAM, JACK B. and MEREDITH F. BURRILL
  1956. Water for Industry. Pub. No. 45, AAAS Symposium, 1953. American Association for the Advancement of Science. Washington, D.C. 131 pp.
GREAT LAKES WATER QUALITY, REPORT ON
  1980. J. McGuire (U.S.) and R. W. Slater, (Canada), Chairs. International Joint Commission, Great Lakes Regional Office. Windsor, Ontario. 68 pp.
IDAHO ENVIRONMENTAL ISSUES
  1974. Idaho Wildlife Federation, Environmental Education Fund. Boise, Idaho. 100 pp.
JACOBS, LEE W., ed.
  1977. Utilizing Municipal Sewage Wastewaters and Sludges on Land for Agricultural Production. North Central Regional Extension Publication No. 52. Michigan State University. East Lansing, Mich. 75 pp.
KNEEBONE, WILLIAM R., and IAN L. PEPPER
  1982. Consumptive Water Use by Subirrigated Turfgrasses under Desert Conditions. Agron. J. Vol. 74. May-June. pp. 419–423.
KNIGHT, JR., LEAVITT A.
  1973. The Purity and Impurity of Our Tap Water. The American Legion Magazine. Vol. 94, No. 1. New York, N.Y. pp. 12–17, 46–52.

KOTTER, CLEON M.
   1981. Surge Flow Concepts Cause Big Ripples in Surface Irrigation Research in Utah. Irrigation Age. Vol. 15, No. 6. March. St. Paul, Minn. pp. 22–23.

LARSON, RON
   1981. Moving Water, Conservation Offered as Solutions to Water Problems. Irrigation Age. Vol. 16, No. 1. St. Paul, Minn. pp. 62–63.

OCEANS AND CLIMATE
   1978. William H. MacLeish (ed.). Oceanus. Vol. 21, No. 4. Woods Hole Oceanographic Institution. Woods Hole, Mass. 71 pp.

PLANNING AND BUILDING THE GOLF COURSE
   1981. Killian and Assoc. (eds.). National Golf Foundation. North Palm Beach, Fla. 48 pp.

PROBLEMS RELATING TO SAFE WATER SUPPLY IN SOUTHEASTERN, MINNESOTA
   1976. Report to the Legislative Committee on Minnesota Resources. Minnesota Department of Health. St. Paul, Minn. 85 pp.

THORNE, WYNNE
   1963. Ed., Publ. No. 73 AAAS Symposium 1961. American Association for the Advancement of Science. Washington, D.C. 364 pp.

WARD, COLEMAN Y.
   1969. Climate and Adaptation. Turfgrass Science. American Society of Agronomy, Monograph 14. ASA, Madison, Wis. pp. 27–38.

WATER AND LAND: MINNEAPOLIS-ST. PAUL FUTURE PERSPECTIVES AND PLANS
   1977. Upper Mississippi River Basin Commission Federal Bldg., Fort Snelling, Minneapolis, Minn. 164 pp.

WATER RESEARCH: A SOUND INVESTMENT
   1982. Lou Ellen Ruesink (ed.). The Water Resources Institute and Texas Agricultural Experiment Station, College Station, Tex. 17 pp.

WATER SUPPLY—WASTEWATER TREATMENT COORDINATION STUDY
   1979. Report to Congress, Public Comment and Review Draft. U.S. Environmental Protection Agency, Office of Drinking Water. Washington, D.C. 375 pp.

WATER: WILL THERE BE ENOUGH?
   1979. Robin Murphy (ed.). Water Foundation. Santa Barbara, Calif. 110 pp.

WEATHER FORECASTING
   1977. UCAR Forums: 1975–1976. National Center for Atmospheric Research. Boulder, Colo. 41 pp.

Practicum

4 Physiology of water use and water stress

VICTOR B. YOUNGNER

Victor B. Youngner received his Ph.D. degree in plant breeding from the University of Minnesota. From 1955 to 1984, Dr. Youngner was an agronomist with the University of California, first at UCLA and later at the Riverside campus. Dr. Youngner died on April 18, 1984 at Riverside.

An understanding of how a plant uses water and how it responds to water shortages is essential if water-efficient turfgrasses and efficient water-use programs are to be developed.

The movement of water upward in plants results from the concentration of dissolved solids in water in plant tissue. When there are two different levels of concentration of these solids, water tends to move from the area of low concentration and high free water to the area of high concentration and low free water. This happens between soil and roots, among plant cells, and between plants and air. This is one of the major forces for water transport within the plant—a force so powerful that it is capable of lifting water to the tops of the tallest trees.

Water enters a plant through the root hairs near the root tip and diffuses into the water-conducting cells. The upward movement of water carries it through the roots, stem, and leaves out into the atmosphere. Water leaves the plant through the stomatal pores of the leaves in a process called *transpiration*. The term *evapotranspiration* refers to the combined water loss of evaporation from the soil and transpiration from the plant.

When the amount of water available in plant tissue for growth and evapotranspiration drops below a certain point, a water deficit occurs, creating water stress. On bright, sunny days, transpiration can be faster than water absorption. When that happens, leaves, stems, and roots may have water deficits even in moist soil.

Wilt is the loss of turgor (internal pressure) within plant tissue. The turf first shows a patina of gray-blue color. The grass blades do not spring back when stepped on and they appear dry and limp. In the early stages, an application of water leads to satisfactory recovery. If the permanent wilting point is reached, the turf will not recover and the plants die. When drought occurs, moisture in the root zone may be below the permanent wilting point.

Plants adapt to resist the effects of drought through the development of extensive root systems, thickened cuticles, rolled leaves, and stomatal closure. Reducing absorption of radiant heat from the sun through adaptations, such as leaf hairs and steeply inclined leaves, also reduces water loss, thus reducing stress.

Some turf species have developed mechanisms to resist drought by escaping its effects. The truly annual strains of *Poa annua* exhibit seed dormancy and are able to complete the life cycle in a few weeks, perhaps before the warm, dry summer. Perennials like buffalograss, *Buchloe dactyloides* (Nutt.) Engelm.; bermudagrass, *Cynodon dactylon* (L.) Pers.; and kikuyugrass, *Pennisetum clandestinum* Hochst. ex Chiov., survive drought periods through rhizomes that survive even in dry soil.

When water stress occurs, an early response of the grass plant is to close its stomata to reduce water loss; when the stress abates, the stomata may reopen. As a result of closure, growth slows from the depression of photosynthesis caused by the decrease in transpiration.

Continued water stress also results in reduced tillering and, in time, a loss of turf density. Shorter rhizomes and stolons with shorter internodes will also be produced.

Root growth will not be affected as readily, leading to an increase in the root:shoot ratio. Moderate water stress may actually stimulate deeper rooting. By using this fact and gradually lengthening the intervals between irrigations to create gradual water stress, irrigators may produce turf better able to withstand drought later.

There is frequently a surge of growth noted at the end of a droughty period. Plants under water stress accumulate soluble carbohydrates in leaves, roots, and stem bases. The availability of these stored carbohydrates may be an important factor in the growth surge.

Grasses have been divided into warm-season and cool-season grasses, essentially based upon climatic adaptability. To be technically correct, the specific groups are actually based on the molecular structure of a product of photosynthesis; however, the climatic reference is commonly accepted. Warm-season grasses are usually subtropical species found in the Sunbelt. Cool-season grasses are usually found in areas that have cold winters. Both grasses tolerate relatively moderate climates and can be found growing side by side, providing acceptable turf in the transition climate zones.

Warm-season grasses generally have lower water-use rates. Some warm-season grasses, such as bermudagrass and zoysiagrass (*Zoysia japonica* Steud.), have well-developed mechanisms for drought avoidance, primarily an extensive root system. In addition, these two species have the ability to tolerate high concentrations of solutes in internal water.

Some cool-season turfgrasses, primarily the fescues, have leaf characteristics that are effective in limiting transpirational water loss. Leaves of the red fescue group (*Festuca rubra*) are normally rolled somewhat lengthwise with stomata on the inside. When water deficits occur, the leaves roll more tightly, enclosing the stomata in tiny chambers with high humidity and low evaporation. Tall fescue leaves (*Festuca arundinaceae* Schreb.), though not normally rolled, will roll in a similar manner under water stress. Tall fescue has a fairly high water-use rate, but good drought tolerance results from its extensive root system.

The knowledge of how a plant responds to water stress is valuable in managing turf. Recognition of wilt and drought symptoms and the effects on the grasses is vital to successfully growing turf.

—*Prepared by Stephen T. Cockerham*

4 Physiology of water use and water stress

Victor B. Youngner

Plant-water relations have intrigued plant physiologists for many years, and the realization has been growing that the world's supply of fresh water is not inexhaustible. In many parts where demand far exceeds supply, study has moved from the purely academic to one involving applied scientists and technologists in many fields. An understanding of how a plant uses water and how it responds to water shortage is essential to development of water-efficient turfgrasses and efficient water-use strategies. A large body of literature has now accrued on the subject. The purpose of this discussion is to examine broadly a few significant concepts that are relevant to water conservation in turfgrass culture.

As in any field, special terminology to describe various aspects of water relations has developed. To communicate ideas and research results to scientists and the public, a general understanding of these terms and concepts is necessary.

Water potential

The water status of the soil-plant-atmosphere continuum is generally described now in units of water potential, a term adopted from the chemical potential or free energy of water within the soil or plant (21). Water potential is expressed as the difference between the chemical potential of pure free water, assigned a water potential of 0 bars, and the chemical potential of water within a soil-plant-atmosphere system. Because the chemical potential of water within the system is usually lower than that of pure free water, it is expressed in negative water-potential values. In many publications, water potential is now expressed as megapascals instead of bars; 1 megapascal equals 10 bars.

The three principal components of plant-water potential are matric, osmotic, and pressure potentials. Matric potential results from forces of capillarity, hydration, and adsorption associated with cell walls and colloidal surfaces within the plant. These reduce the free energy of water below that of pure free water; thus, matric potential has negative values. Osmotic potential results from the presence of dissolved solutes which reduce the free energy below that of pure free water so that it, too, has negative values. Pressure potential is due to turgor pressure acting outwardly against the cell walls and internal membranes of the plant. Turgor pressure results from forces of hydrostatic pressure in excess of atmospheric pressure. It is generally positive, except when the plant is wilting.

Although other elements may be involved to various degrees, in a simplified form plant-water potential is considered to be the sum of osmotic, matric, and pressure potentials and may be written in the form of the equation:

$$\psi = \psi_\pi + \psi_\gamma + \psi_\rho$$

In which:

ψ = water potential
ψ_π = osmotic potential
ψ_γ = matric potential
ψ_ρ = pressure potential

Water movement in the soil-plant-atmosphere continuum is always along a gradient from higher to lower free energy, or in terms of water potential, from a higher water potential to a lower. A gradient of declining water potential usually exists from the soil, through the plant's vascular system, to the stomata and into the atmosphere. This gradient is one of the major forces for water transport within the plant.

Water movement in unstressed plants

Water enters a plant through absorption by the roots, primarily the root hairs near the root tip, and diffuses through the root cortex into the vascular system, the water-conducting cells of the xylem. Absorption and diffusion of water into the root result from a declining water potential from soil to root hair to cortex to xylem. The upward movement of water in the xylem carries it through the root stem and leaves and out into the atmosphere as water vapor through the stomatal pores of the leaves. (See figure.) A minute amount may also escape through the cuticle. In the plant, water exists in the liquid phase until it reaches the substomatal cavity where it changes to the vapor phase. Loss of water through the leaves is called transpiration, and the upward translocation from roots to leaves is called the transpirational stream, as it is powered by the transpiration process. The transpirational pull that lifts the column of water through the plant results from adhesion of the water molecules to the walls of the xylem cells and the cohesion of the molecules to each other. It is a powerful force capable of lifting water to the tops of the tallest trees. Dissolved mineral nutrients are carried in the transpirational stream to sites where they are utilized for plant growth.

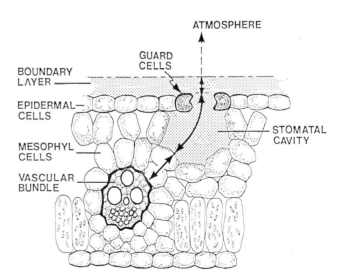

The movement of water vapor from the mesophyll evaporative surfaces to the upper air mass and the resistances to this movement
—*Drawing courtesy S. M. Batten and J. B. Beard, Texas A & M University, College Station*

Transpiration may appear to be wasteful, as only about 1 percent or less of the water absorbed by the roots is actually used in metabolic processes by most higher plants. However, water is also used to transport mineral nutrients, metabolites, and products of photosynthesis. Evaporation from the leaf is an important cooling process.

The stomatal apparatus of grasses consists of two dumbbell-shaped guard cells which function to open and close the stomatal pore to regulate gas exchange. Stomatal pores are the main passageways for the inward diffusion of carbon dioxide for photosynthesis and the outward diffusion of water vapor in the transpirational process. On each side of the guard cells are two subsidiary cells of various shapes, depending upon species, that are thought to assist in the opening and closing of the stomata. Resistance to outward diffusion of water vapor depends on shape and size of the substomatal cavity and the stomatal pore, and the number of stomata per square centimeter of leaf surface (13). There is also an external resistance due to the thin layer of moist, unstirred air surrounding the leaf. In most plant communities, boundary-layer resistance is considered unimportant, but in a dense turf with little air movement, it may be substantial.

Water-use rate

Water-use rate or consumptive water use refers to the amount required by a stand of grass for growth and evapotranspiration. Water-use rates change with the season as they are affected by climatic conditions that affect evapotranspiration rates and that differ for different species (23). Water-use rates are also affected by soil-moisture availability and cultural practices (5, 12). When comparative water-use rates are given for a grass or turf, sufficient soil moisture is assumed available to meet all needs for growth and evapotranspiration unless otherwise specified. Water-use rates may be presented on a daily, weekly, monthly, or yearly basis.

Water deficit and water stress

Water deficit is a term variously understood and used. Correctly used, it refers to reduction in the quantity of water in a plant tissue compared with the quantity present at a selected reference base, usually a water potential of 0 bars. A water deficit then may be thought to exist any time the water potential is less than 0 bars. A plant, even though it is growing in moist soil, will experience daily periods of water deficit. On hot sunny days, water absorption may lag behind transpiration so that leaves, stems, and even roots may have water deficits. Water deficits of this nature, if not prolonged, may have little or no effect on plant growth, although some temporary wilting may occur.

When a grass plant experiences more severe and prolonged water deficits in its tissues, growth and physiological activity within the plant are adversely affected. Insufficient soil moisture is the stress factor, and the plant is said to be suffering from water-deficiency stress or, simply, water stress; that is, injurious effects of the water shortage. Plants will exhibit some symptoms of water stress when leaf water potential drops to low levels.

When leaf-water potentials remain at low levels for long periods, plant survival becomes a question. Species and varieties differ in their ability to tolerate water stress.

Drought and drought resistance

Drought, usually thought of in terms of weather or soil moisture, may be said to exist when there has been a prolonged period without rainfall or irrigation so that moisture in the root zone is depleted. Wilting, which develops from loss of turgor, results from weather conditions, osmotic characteristics, and root density, as well as soil moisture conditions (20). Drought is not a particularly good scientific concept, but it is a useful term to indicate a condition wherein a plant experiences severe water stress.

Plants differ not only in their ability to resist severe water stress but also in the ways they resist drought. Drought resistance is the general term covering all adaptations that permit plants to survive periods of water stress. Three types of drought resistance are recognized: avoidance, tolerance, and escape.

Avoidance. Many plants resist water stress by avoidance (14). They are able to maintain high internal water potentials even when subjected to drought. This is accomplished by various anatomical and morphological adaptations that maintain adequate water uptake or reduce water loss. Examples of drought avoidance are: development of root systems able to explore the soil more extensively for moisture as drought develops, a high root-to-shoot ratio, development of xeromorphic structures such as a thick cuticle or rolled leaves, and stomatal closure to reduce transpirational water loss. Reduction of radiation absorption through adaptations such as steeply inclined leaves, reflective leaf hairs, and thick wax coatings on leaves also reduce water loss.

Tolerance. On the other hand, plants exhibiting drought tolerance have adaptations whereby they are able to tolerate low internal water potentials (14). Tolerance may be achieved through maintenance of turgor pressure as plant water potential declines or through the ability of the plant's protoplasm to survive severe desiccation. Although tolerance may exist to some degree in many crop plants, including the grasses, it may be of lesser general importance in agriculture than avoidance. Tolerance mechanisms may only permit the plant, if under severe drought, to survive; avoidance also allows growth and development to occur. Many plants possess both avoidance and tolerance to some degree. For turfgrasses, avoidance appears to be the most important form of resistance.

Escape. Drought escape, a third form of resistance, is sometimes considered a method of drought avoidance. It may take the form of an ability to complete the life cycle during a brief rainy season before soil moisture is depleted, or it may be as some type of dormancy. Many weedy turf plants, such as *Poa annua* L., are able to complete their life cycles in a few weeks and thus may be considered drought escapers even though they require high soil moisture during the growth period.

Dormancy of seeds due to endogenous growth inhibitors, which prevent germination until a period of favorable soil moisture, may permit drought escape. The truly annual strains of *Poa annua* exhibit seed dormancy while the more perennial types do not (18). Perennials such as buffalograss (*Buchloa dactyloides* [Nutt] Engelm.), bermudagrass (*Cynodon dactylon* [L.] Pers.), and kikuyugrass (*Pennisetum clandestinum* Hochst. ex Chior.) survive drought periods through rhizomes which survive even in dry soil (10).

Efficiency of water use

Water-use efficiency is an important concept in crop production (3), but it is less so in turf culture. Water-use efficiency is defined as units of dry weight produced per unit of water transpired. Several units may be used, such as grams of dry matter per kilogram of water or tons per acre-inch. It should not be confused with drought resistance where emphasis is on survival. Although dry matter production is not the major objective in turf culture, grass growth is necessary to have a good turf so selection of species and varieties that are efficient water users may be a desirable goal.

A need may exist to develop a similar concept for turf culture employing parameters more appropriate than dry matter yield. Scales for criteria of turf quality, such as texture, color, density, or playability and wear tolerance in the case of sports turfs, could be established and related to water use. It seems evident that turfgrass species differ in their ability to maintain quality per unit of water transpired just as they are in their ability to produce dry matter.

General responses to water deficits

When water deficits occur, an early response of the grass plant is stomatal closure which reduces water loss. Stomatal closure is a major plant response to water stress. Closure may occur at the time of peak water demand when temperatures are high or winds are blowing. When this environmental stress abates, stomata may reopen (15). Although stomatal closure in response to temporary water deficits may occur daily and cause little effect on plant growth, if the water deficit becomes greater and the plant is subjected to more severe water stress, stomata will remain closed for longer periods.

Stomatal closure may occur at a critical threshold leaf-water potential or relative leaf-water content and will be little affected until that potential is reached. Complete closure of stomata may take place over a very narrow range of leaf-water potential (3). The exact potential for stomatal closure varies with age of plant and leaf, position of the leaf in the canopy, history of previous stress, and perhaps species or variety. Differences in stomatal response may determine a species' ability to tolerate severe water stress. Stomata of many species may close in response to low atmospheric humidity even in the absence of a leaf-water deficit.

When stomata close under water stress, CO_2 diffusion into the leaf is affected as is the outward diffusion of water vapor. Consequently, there is a depression of photosynthesis corresponding to the decrease in transpiration with stomatal closure (7). Although stomatal closure has been clearly shown to be a primary cause of photosynthesis reduction under water stress, reduction due to depression of chloroplast and enzyme activity also occurs under prolonged stress (6, 4).

Depression of photosynthesis will obviously retard growth (8). One of the first effects of water stress is a retardation of leaf expansion due to a decrease in cell size. Leaf expansion is highly sensitive to water stress and may occur before stomatal closure and reduction of photosynthesis. As the stress continues or deepens, rate of cell division will also be reduced, contributing to further decrease in leaf size. When soil moisture is again provided, leaf expansion by increase in cell size resumes almost immediately while there is a lag in resumption of photosynthesis.

Continued water stress also affects other aspects of grass shoot growth. In addition to reduction in leaf size, there is a retardation in the rate of leaf emergence due to reduced meristematic activity of the shoot apex as well as decreased rate of cell enlargement. Reduction of meristematic activity also leads to a reduced tillering rate and, consequently in time, a loss of turf density. Lowered activity of intercalary meristems and decreased rates of cell enlargement also produce shorter rhizomes and stolons with shorter internodes.

Root growth, on the other hand, is not affected as much by water deficits, leading to an increase in the root:shoot ratio. Moderate water stress may actually stimulate root extension into deeper soil strata with more abundant moisture, further increasing the root:shoot ratio. Return to an adequate supply of soil moisture will decrease the ratio by inducing increased top growth.

The decrease in top growth, increase in root growth, and resulting increase in root:shoot ratio is an important method of drought avoidance. Reduction of leaf-surface area reduces the rate of water loss through transpiration while the increased root system improves the plant's ability to take up water from moist soil at greater depth. In some regions, these responses may be used to advantage in preparing for anticipated water shortages. Gradual lengthening of the interval between irrigations to impart a gradual water stress may produce a turf better able to withstand more severe stress later.

Effects of water deficits on physiological processes

Water deficits have been shown to affect other physiological processes within the plant besides photosynthesis. Despite the reduction in carbon assimilation, plants under water stress often accumulate soluble carbohydrates in leaves, roots, and stem bases (11, 16). The cause of this carbohydrate accumulation is not clear, but it may be associated with the great sensitivity of shoot growth to water stress previously mentioned. Mineral deficiency and reduced enzyme activity brought about by reduced water absorption also may be involved. Growth reduction before photosynthesis is stopped by stomatal closure could lead to increased carbohydrate storage.

The availability of these stored carbohydrates may be important to the surge of growth noted in many plants at the end of a drought period. With the return of adequate soil moisture, stored carbohydrates form a readily available source of energy for renewed growth. Reduction in stored carbohydrates, associated with regrowth following cessation of drought, has been observed (16). Thus, carbohydrate accumulation during drought provides a means for rapid recovery at its termination.

Biochemical changes. Numerous biochemical changes occur in plants subjected to water stress. The significance of many of these changes to drought tolerance is not clear at present. A greatly reduced nitrate reductase activity in water-stressed plants may be related to reduced nitrogen accumulation and reduced dry matter production (19). Although there is considerable variation among species, protein synthesis is generally reduced when plants are subjected to water stress. In many plants moderately tolerant of water stress, protein synthesis is resumed upon resupply of water if stress has been moderate. However, if stress has been prolonged, the ability to resume protein synthesis appears to be lost (9).

Certain amino acids accumulate in water-stressed plants (2). Of these, proline accumulates most consistently at a fairly rapid rate across a wide range of species. A correlation between proline accumulation and drought resistance has been postulated (14). Proline concentration may also be related to the plant's ability to recover upon water-stress relief.

Osmotic adjustment. An important mechanism for water-stress tolerance is osmotic adjustment. As plants are stressed, bringing about a loss of water from the cells, solutes are concentrated within them leading to decreased osmotic potential. Solutes may also be translocated to these cells to produce a further decrease in osmotic potential (17). Several different solutes may enter into the osmotic adjustment process, but soluble carbohydrates and free amino acids may be of primary importance. The rapid accumulation of large amounts of proline indicates that it may be of special significance (1).

Many plants show a decrease in the threshold critical water potential for stomatal closure following cycles of water stress (10) which can best be explained by osmotic adjustment. Through osmotic adjustment, permitting stomata to stay open longer, plants are able to continue more active photosynthesis during periods of water stress, thus at least partially alleviating stress damage.

Leaf distinctions. Grasses, as well as other higher plants, have been divided into two groups on the basis of the number of carbon atoms in the molecule of the first carboxylation product of photosynthesis. These are conveniently termed C_3 and C_4 species (72). A distinctive leaf anatomy of C_4 species provides an easy method for distinguishing members of the two groups. Most warm-season grasses have the C_4 photosynthetic pathway while most cool-season grasses have the C_3 pathway. Grasses possessing the C_4 pathway in general have a lower CO_2 compensation point, require more light to saturate photosynthesis, and have a higher optimum temperature for growth. These characteristics may account for the generally higher water-use efficiency and lower water-use rate of the C_4 species (see Table), especially in regions of high solar energy or during the warm season.

Comparative peak water-use rate (WUR) and water-use efficiency (WUE) of C_3 and C_4 grasses*

Rates	C_3	C_4
WUR (mm/day)	5.5	3.7
WUE (g dry wt/kg water)	1.49	3.14

*Water-use rate adapted from data (22); water-use efficiency data (3). Specific amounts very with species, location, and environmental conditions.

Some cool-season turfgrasses, primarily the fescues, have leaf characteristics that help limit transpirational water loss. Leaves of the red fescue group are normally rolled somewhat lengthwise with stomata on the inside. When water deficits occur, the leaves roll tighter, enclosing the stomata in tiny chambers with high humidity. Tall fescues, though not normally rolled, will roll in a similar manner under water stress.

Other factors. Atmospheric water stress occurs when weather conditions bring about excessive transpirational water loss even though adequate levels of moisture are in the soil. A "wet wilt," which may be as harmful as any other water stress, may result. Examples of the more common weather conditions creating this problem are extremely high temperature, low humidity, and "Santa Ana"-type winds. This is a frequent problem on such shallow-rooted turfs as bentgrass greens. Plants may also wilt when the soil is saturated because low soil oxygen levels inhibit the root system's ability to absorb moisture.

A similar and equally harmful kind of water stress may occur in winter during bright sunny days. The leaves and the air immediately around them may become quite warm while the soil remains cool. The roots in the cool soil are unable to take up sufficient water to meet transpiration water loss. Under these conditions, leaves may wilt and extreme desiccation of the plant may occur.

Summary and conclusion

The concepts discussed are based largely on studies of plants other than the turfgrasses. While their application to turfgrasses and turfgrass culture is in little doubt, there is almost no data specific to this group of plants.

Turfgrasses differ in water-use rates, responses to water deficits, and in resistance to severe water stress. Some turfgrasses, such as bermudagrass (*Cynodon dactylon* [L.] Pers.) and zoysia (*Zoysia japonica* Steud.), have well-developed mechanisms for drought avoidance, due primarily to their extensive root systems, and they may have adaptations for drought tolerance as well.

Although these two species also have a moderate-to-low water-use rate, drought resistance and low water-use rate do not necessarily go hand in hand. Tall fescue (*Festuca arundinacea* Schreb.), for example, has a fairly high water-use rate but good drought resistance.

Water conservation in turfgrass management is a critical necessity. Turfgrass breeders, using a knowledge of the concepts and principles discussed in this and other chapters, can develop better turfgrasses with low water-use rates and high drought resistance.

Literature cited

1. ASPINALL, D., and L. G. PALEG
   1981. Proline accumulation: physiological aspects. pp. 205–259. *In* L. G. Paleg and D. Aspinall (eds.), The Physiology and Biochemistry of Drought Resistance in Plants. Academic Press, N.Y.
2. BARNETT, N. M., and A. W. NAYLOR
   1966. Amino acid and protein synthesis in bermudagrass during water stress. Plant Phys. 41:1222–1230.
3. BEGG, J. E., and N. C. TURNER
   1976, Crop water deficits. pp. 161–217. *In* N. C. Brady (ed.), Advances in Agronomy. Academic Press, N.Y.
4. BERKOWITZ, G. A., and M. GIBBS
   1982. Effect of osmotic stress on photosynthesis studied with isolated spinach chloroplasts. Site-specific inhibition of the photosynthetic carbon reduction cycle. Plant Phys. 70:1535–1540.
5. BIRAN, I., B. BRAVDO, I. BUSHKIN-HARAV, and E. RAWITZ
   1981. Water consumption and growth rate of 11 turfgrasses as affected by mowing height, irrigation frequency, and soil moisture. Agron. J. 73:85–90.
6. BOYER, J. S.
   1971. Recovery of photosynthesis in sunflower after a period of low leaf water potential. Plant Phys. 47:816–820.
7. CRAFTS, A. S.
   1968. Water deficits and physiological processes. pp. 85–133. *In* T. T. Kozlowski (ed.), Water Deficits and Plant Growth II. Academic Press, N.Y.
8. GATES, C. T.
   1968. Water deficits and growth of herbaceous plants. pp. 135–190. *In* T. T. Kozlowski (ed.), Water Deficits and Plant Growth II. Academic Press, N.Y.
9. HSIAO, T.C.
   1970. Rapid changes in the level of polyribosomes in *Zea mays* in response to water stress. Plant Phys. 46:281–285.
10. JONES, M. M., N. C. TURNER, and C. B. OSMOND
    1981. Mechanisms of drought resistance. pp. 15–35. *In* L. G. Paleg and D. Aspinall (eds.), The Physiology and Biochemistry of Drought Resistance in Plants. Academic Press, N.Y.
11. JULANDER, O.
    1945. Drought resistance in range and pasture grasses. Plant Phys. 20:573–599.
12. KNEEBONE, W. R., and I. L. PEPPER
    1982. Consumptive water use and subirrigated turfgrasses under desert conditions. Agron. J. 74:419–423.
13. KRIEDMANN, P. E., and W. J. S. DOWNTON
    1981. Photosynthesis. pp. 283–314. *In* L. G. Paleg and D. Aspinall (eds.), The Physiology and Biochemistry of Drought Resistance in Plants. Academic Press, N.Y.
14. LEVITT, J.
    1980. Responses of plants to environmental stresses. Vol. II. Academic Press, N.Y. 607 pp.
15. MANSFIELD, T. A., and W. J. DAVIS
    1981. Stomata and stomatal mechanisms. pp. 315–346. *In* L. G. Paleg and D. Aspinall (eds.), The Physiology and Biochemistry of Drought Resistance in Plants. Academic Press, N.Y.
16. MELA, T., and V. B. YOUNGNER
    1976. Recovery of three temperature-climate grasses from drought stress. Annales Agriculture Fenniae 15:309–315.
17. MEYER, R. F., and J. S. BOYER
    1972. Sensitivity of cell division and cell elongation to low water potentials in soybean hypocotyls. Planta 108:77–87.
18. SGAMBATTI, LUCIANO
    1978. Germination and dormancy studies in *Poa annua* L. Ph.D. thesis. Univ. of Calif., Riverside.
19. SINHA, S. K., and D. J. D. NICHOLS
    1981. Nitrate reductase. pp. 145–169. *In* L. G. Paleg and D. Aspinall (eds.), The Physiology and Biochemistry of Drought Resistance in Plants. Academic Press, N.Y.
20. SLATYER, R. O.
    1957. The significance of the permanent wilting percentage in studies of plant and soil water relations. Bot. Rev. 23:858:636.
21. SLATYER, R. O., and S. A. TAYLOR
    1960. Terminology in plant- and soil-water relations. Nature 187:922–924.
22. WALKER, S. S., and J. K. LEWIS
    1979. Occurrence of C_3 and C_4 photosynthetic pathways in North American grasses. J. Range Manag. 32:12–28.
23. YOUNGNER, V. B., A. W. MARSH, R. A. STROHMAN, V. A. GIBEAULT, and S. SPAULDING
    1981. Water use and turf quality of warm-season and cool-season turfgrasses. pp. 251–257. *In* R. W. Sheard (ed.), Proc. Fourth International Turfgrass Research Conf., Univ. of Guelph, Guelph, Ontario, Canada.

Practicum

5 An assessment of water use by turfgrasses

JAMES B. BEARD

James B. Beard is professor of turfgrass physiology and ecology, Department of Soil and Crop Sciences, Texas A & M University. His research program emphasizes stress physiology in perennial grasses, including the dimensions of water conservation, drought resistance, wear tolerance, shade adaptation, heat tolerance, and winter injury, plus such cultural dimensions as minimal maintenance turf culture and sod production.

Water-use rate is the total amount of water required for turfgrass growth plus the quantity transpired from the grass plant and evaporated from associated soil surfaces. It is typically measured as evapotranspiration and expressed as ET in inches per day.

The comparative water-use rates of turfgrass species are distinctly different from the relative drought tolerance because each is a distinctly different physiological phenomenon. For example, tall fescue is one of the more drought-tolerant cool-season turfgrasses, but it possesses a very high water-use rate.

Reducing the turfgrass water-use rate is a strategy associated with irrigated turfs. The goal is to select turfgrasses that require the least possible supplemental water via irrigation.

There are major differences in the comparative water-use rates among the 25 turfgrass species used throughout North America. The most definitive work has been done with the warm-season turfgrasses—those species adapted to soil temperatures in the 80° to 95°F range and commonly grown in the southern half of the United States. The high-density, low-growing cultivars of zoysiagrass and bermudagrass, plus buffalograss and centipedegrass, have exhibited low water-use rates. Other warm-season species, such as St. Augustinegrass, seashore paspalum, bahiagrass, and kikuyugrass, have exhibited relatively high water-use rates.

Among the cool-season species, which grow best at soil temperatures in the 60° to 75°F range, comparative information is limited. However, there are indications that Kentucky bluegrass ranks somewhat low in water-use rate; tall fescue, perennial ryegrass, and creeping bentgrass have exhibited relatively high water-use rates. No data are available on the fine-leafed fescues and other cool-season turfgrass species used as frequently.

The professional turf manager and homeowner should be aware of the particular plant characteristics that contribute to a low water-use rate. The major factors are a slow vertical leaf extension rate and a high canopy resistance, the components of which include:

- High shoot density,
- Narrow leaf, and
- More horizontal leaf orientation.

5 An assessment of water use by turfgrasses

James B. Beard

A key component in a water conservation strategy is the selection of turfgrass species and cultivars possessing a low water-use rate. In the past, when water was readily available at low cost, little attention was paid to water conservation strategies. This situation, combined with a need for research in many diverse areas of turfgrass culture, resulted in very little information being generated concerning water-use rates of turfgrasses (1). However, the general public has now become aware of the developing water problem and is beginning to support research into turfgrass water conservation. A limited number of studies were completed in the 1970s and a major expansion in these efforts has occurred during the early 1980s. Our current state of knowledge will be addressed in this chapter. It will become apparent that major voids still exist in our basic pool of knowledge regarding turfgrass water use. It is hoped that the limited state of our knowledge will be substantially reduced during this decade.

Plant/water relations

A properly functioning turfgrass community requires water for survival and growth. The typical turfgrass plant has a water content that ranges from 75 to 85 percent by weight (1). A 10 percent reduction in water content from 75 to 65 percent within a short time may be lethal.

Only a very small portion of the water absorbed by a turfgrass plant is actually utilized in physiological and growth processes. By far the major portion of this water is transpired from the turfgrass surfaces into the surrounding atmosphere. Sometimes water that flows through the plant is considered a loss. There is now a substantial body of research to prove that transpiration is beneficial. In the case of turfs, transpiration is important to survival during heat stress due to cooling of the aboveground tissues.

Water use by turfgrasses is a dynamic system involving interactions among the soil, the turfgrass plant, and the surrounding atmosphere. The flow of water follows a sink-source concept with the soil being the primary source of water and the atmosphere being its ultimate sink. This flow of water from the soil through the plant into the atmosphere is driven by a gradient in the free energy of water. Typically, there is a gradient of decreasing free energy from the soil to the plant through the vascular system of the stems to the leaves and through the leaves to the atmosphere. Distribution and movement of water through the soil, plant, and atmosphere may occur in liquid or gaseous phases. Water may occur in soils as liquid or vapor; in the plant it occurs as a liquid continuum extending throughout the vascular system. There is a phase change in the substomatal cavity from liquid to vapor that results in transpiration of water from the leaf to the atmosphere. Thus, turfgrass water relations are a complex system closely interlinked by an ever-changing energy continuum.

Evapotranspiration

Key to the flow of water from the soil source to the atmosphere is evaporation (E), a process that may occur via two paths involving movement from the soil to the atmosphere or from the soil into the plant and to the atmosphere. Thus, there are both plant and soil components to the evaporative surfaces. The term transpiration (T) relates to evaporation from plant surfaces only. The term evapotranspiration (ET) combines the evaporative processes that occur from the soil (E) and by transpiration from the plants growing thereon (T).

The measurement and mathematical derivation of ET are summarized in the section of this chapter on resistances to evapotranspiration. ET is commonly expressed quantitatively as milligrams per square meter per second (mg m^{-2}s^{-1}) or as millimeters per day (mm day^{-1}). A classification of evapotranspiration rates for turfgrasses is presented in Table 1, which should provide a standardized quantitative base reference when using the more generalized relative terms such as a high or a low ET rate.

The term water-use rate (WUR) refers to the total amount of water required for turfgrass growth plus the quantity transpired from the grass plants and evaporated from the associated soil surfaces, expressed quantitatively per unit of time.

$$WUR = \frac{W_p + W_t + W_e}{t}$$

WUR - water-use rate
t - time
W_p - water used internally by plants for growth
W_t - water transpired from plants
W_e - water evaporated from soil

The practitioner typically expresses WUR as inches per week (in/wk); researchers express it as millimeters per day (mm day^{-1}).

Another comparable term occasionally used is consump-

TABLE 1 A classification of evapotranspiration rates for turfgrasses

Relative ranking	Evapotranspiration rate	
	mm day^{-1}	inches/week
Very low	< 4.0	< 1.0
Low	4.0 to 4.9	1.1 to 1.3
Medium-low	5.0 to 5.9	1.4 to 1.6
Medium	6.0 to 6.9	1.7 to 1.9
Medium-high	7.0 to 7.9	2.0 to 2.2
High	8.0 to 8.9	2.3 to 2.5
Very high	> 9.0	> 2.5

tive use. However, water-use rate is preferred, as it is more descriptive and permits easy information transfer between the scientist and practitioner in the field. Another term frequently used in crop and horticulture literature that is confused with water-use rate is water-use efficiency (WUE). This is defined as the amount of dry matter produced per unit mass of water lost and thus is distinctly different from water-use rate. Water-use rate is the appropriate terminology for use in turfgrass science, since functional survival of the grass plant is the important criterion rather than its yield.

One final aspect that sometimes leads to confusion, especially by the field practitioner, is that a grass possessing a low water-use rate is not necessarily drought tolerant or resistant. Each characteristic has a distinctly different physiological basis. However, a turfgrass species or cultivar that possesses a lower water-use rate can delay onset of drought stress.

In this chapter, the term evapotranspiration (ET) will be used, rather than water-use rate (WUR), because ET is the process quantitatively measured in most studies. Furthermore, WUR is only slightly greater than ET in absolute terms and the differential has no practical effect on relative comparisons of water use among turfgrass species or cultivars. Thus, the term ET, as used herein, is comparable to WUR.

The major thrust of this chapter will be concerned with comparative ET rates at the interspecies and intraspecies levels, as expressed on a daily basis. However, one should be cognizant of seasonal patterns in ET rates as determined by the growth cycle of a turfgrass species, the comparative evaporative demands of the specific climatic region, and the variation in soil-water availability at specific sites. For example, the ET pattern for a warm-season turfgrass in winter dormancy 4 to 5 months out of the year differs distinctly from a cool-season turfgrass growing in the upper portion of the transition zone, where the growing season is 8 to 9 months, or from a cool-season turfgrass in the northernmost cool climatic region, where the growing season is 5 months or less. These aspects of total annual water use must be addressed for each climatic region in relation to each turfgrass species grown there.

One final note: In addressing the issue of water-use rates, the primary concern is with those turfs that are irrigated. The reader should remember that drought resistance is a phenomenon entirely different from water use. It is possible, for example, to have a turfgrass with a medium-to-high water-use rate and good drought resistance. These concepts should be clearly distinguished in relation to the issues addressed in this chapter.

Evapotranspiration measurement

Three basic techniques for the determination of evapotranspiration include the water-balance, energy-balance, and vapor-flow methods (18). The bulk of the data reported in this chapter has been derived from variations of the water-balance method, although an energy-balance method specifically designed for use in well-watered turfgrass situations has been developed and successfully demonstrated (9).

Water-balance methods. The water-balance equation is normally written:

$$P - O - U - ET + \Delta W = O$$

In which:

ET = evaporation from plant and soil surfaces
ΔW = change in soil-water storage during a specified monitoring period
P = amount of precipitation
O = amount of runoff
U = amount of water draining beyond the root zone

In this equation ET is derived by difference, with all other elements either measured or estimated. Essentially, water-balance methods employ the evapotranspiration (ET) term in the water balance equation. The two primary water-balance methods utilized involve (a) lysimetry or (b) monitoring changes in soil-water storage under the plant community.

Lysimetry has been consistently accurate for measuring ET and is best utilized where the vegetation is homogeneous and well watered. These criteria fit nicely with the description of an irrigated turf. This means that lysimeters are well suited for monitoring ET from turfgrasses with the water-balance method. The uniformity, high-plant density, and shallow-root systems typical of perennial grasses maintained as turf also enhance their effectiveness. In addition, mini-lysimeters offer the most accurate method for measuring potential ET.

Mini-lysimeters, as small as 8 liters in size, have been successfully used in turfgrass water-balance methods for monitoring potential ET. Use of fritted clay as the root zone in small lysimeters was proposed (24) and successfully utilized under turfgrass conditions by the Texas research group (6, 7, 8, 10, 19). Site selection and installation of the lysimeters should be carefully executed to ensure minimum disturbance of the canopy during periodic removal of the mini-lysimeter itself, as well as to ensure that the turfgrass community and associated cultural practices utilized on the surrounding turf are comparable.

Mini-lysimeters have been successfully used on both warm- and cool-season turfgrasses (4, 5, 6, 7, 8, 10, 19) maintained in the field (fig. 1). Mini-lysimeters also have been modified for use in conjunction with a controlled-environment growth chamber (fig. 2) capable of simulating high evaporative demands (6, 8, 9).

A second water-balance method involves monitoring changes in soil-water storage at different points under the plant community over a specified period. This method for measuring ET has been utilized on turfs to a limited extent (13, 14). The research group of Charles Wendt at Lubbock, Texas, is utilizing neutron probe techniques to monitor changes in soil-water storage under turfgrasses. This approach may be one of the more promising in monitoring ET rates during progressive water stress where the full effects of the soil profile and root depth need to be evaluated.

Energy-balance methods. The transpiration process depends on an energy supply to satisfy the latent heat-demand compo-

FIGURE 1. A typical mini-lysimeter installation in a series of turf field plots (*above*) with a close view of the mini-lysimeter (*below*)

FIGURE 2. A simulation apparatus for controlled ET research on turfgrasses involving a combination of a controlled-environment chamber for simulation of specific field conditions (*above*) and a series of three mini-lysimeters (*closeup below*)

nent. Thus, transpiration is closely linked to the energy balance, expressed as follows:

$$R_n + H + lE + G + aA = O$$

In which:

- R_n = net radiation flux
- H = sensible heat exchange with the atmosphere
- lE = latent heat exchange with the atmosphere
- G = sensible heat exchange with both vegetation and soil
- aA = energy utilized in photosynthesis and respiration processes within plants

Latent heat exchange from the atmosphere (lE) can be calculated from the equation by difference, if the other components of this energy-balance equation are monitored quantitatively. Energy-balance methods for determining ET are based on the latent heat term in the energy balance equation. They are best adapted for use on homogeneous vegetative covers. Turfgrasses fully meet this criterion.

The concept of potential evapotranspiration originally evolved as a more advanced method for predicting ET. Potential evapotranspiration (PET) is defined as the evaporation from an extended surface of short green vegetation that fully shades the soil surface, that exerts little or negligible resistance to the flow of water, and that is maintained under a nonlimiting water condition. This concept applies to turfgrasses grown under a nonlimiting water condition. In fact, it has been confirmed that actual ET rates for St. Augustinegrass turfs are comparable to potential ET rates when maintained under nonlimiting water conditions (6, 8). Furthermore, the potential ET rate from a vegetative canopy can be accurately defined by measuring ambient weather conditions and the aerodynamic nature of the evaporating surface (11, 22). The potential ET concept is reviewed in the following equation (6):

$$E_p = \frac{(\Delta/\gamma) H/\lambda + X_a/R_a}{\Delta/\gamma + 1} \quad (g \cdot m^{-2} \cdot s^{-1})$$

In which:

E_p = potential evapotranspiration (g·m^{-2}·s^{-1})
Δ = rate of change of saturation vapor density with temperature, i.e., $\left(\frac{\partial X°(T)}{\partial T}\right) T = T_a$ (g·m^{-3}·C^{-1})
γ = volumetric heat capacity of air divided by the latent heat of vaporization (g·m^{-3}·C^{-1})
H = sum of energy inputs at surface exclusive of latent and sensible heat (W·m^{-2})
λ = latent heat of vaporization (J·kg^{-1})
X_a^1 = vapor density deficit of the air (g·m^{-3})
R_a = aerodynamic resistance (s·m^{-1})

The aerodynamic resistance term can be expressed as:

$$R_a = \frac{(\ln[(Z-d)/Z_o])^2}{k^2 \mu} \quad (s \cdot m^{-1})$$

In which:

Z = elevation above the canopy surface (m)
Z_o = roughness parameter (m)
k = Von Karman coefficient (0.41)
μ = wind velocity at Z (m·s^{-1})
d = zero plane displacement (m)

To account for the diffusion of water vapor through the interstitial leaf spaces and stomata, a resistance term (R_s) was included in the equation for actual evapotranspiration and written as:

$$E = \frac{(\Delta/\gamma) H/\lambda + X_a^1/R_a}{(\Delta/\lambda) + 1 + R_s/R_a} \quad (g \cdot m^{-2} \cdot s^{-1})$$

An energy-balance method employing the concepts of potential ET was developed and successfully tested (9). The estimate of ET for turfgrasses growing under nonlimiting water conditions was based on the heat transfer number. The procedure involved recording the rate of cooling of an abruptly shaded canopy and then calculating the heat transfer number by a least-squares method. The procedure was tested in an environmental simulation chamber using an infrared thermometer to record surface temperatures of the turf. The heat transfer numbers so obtained were essentially equal to those obtained by standard mini-lysimeter techniques. Subsequently, this technique was successfully utilized under field conditions by these same investigators. Thus, this may be an attractive technique to monitor ET rates rapidly where large numbers of turfgrass plots are involved.

Vapor-flow methods. This technique involves determining the net vertical flow of water vapor in the air layers near the soil surface. It requires delicate instrumentation to measure vapor fluxes rapidly at specific points in time. It has not been used for turfgrass measurements of ET.

Resistances to evapotranspiration

Evapotranspiration occurs by both stomatal and cuticular pathways. However, the stomatal dimension is by far the most important and will be detailed. Evaporation within the stomatal cavity occurs primarily on the mesophyll cell walls that are exposed to intercellar spaces. Water vapor then diffuses along a gradient through the stomatal cavity outward from the leaf, then through the turfgrass canopy, and eventually into the air mass above (fig. 3). Water vapor encounters resistances to outward movement from the leaf to the atmosphere. These resistances to ET can be separated into three elements expressed as:

$$R_t = R_s + R_c + R_a$$

In which:

R_t = total resistance
R_s = aggregate diffusive resistance of the leaf
R_c = resistance to air mass exchange within the canopy
R_a = the turbulent exchange resistance between the canopy and bulk air

Leaf resistance (R_s) has both cuticular and stomatal components. Cuticular resistance is extremely large, compared with stomatal resistance. The dimensions of stomatal resistance include those encountered in the mesophyll cells, in the intercellular spaces, and in the stomatal cavity as the water vapor moves outward through the stomatal pore. There are resistances to air mass exchange within the turfgrass canopy (R_c) involving a combination of molecular diffusion and turbulent eddy movements. This resistance within the turfgrass canopy can be significant. The turbulent exchange resistance (R_a) between the canopy and bulk air is strongly affected by the thickness and density of the boundary layer.

An understanding of the relative magnitudes of the resistance components controlling ET can provide basic concepts for developing turfgrasses with reduced ET rates, as well as for de-

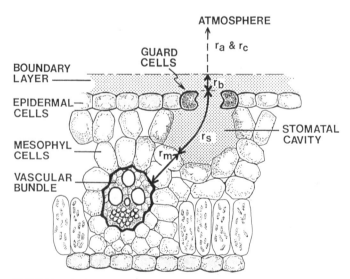

FIGURE 3. An illustration of the path of water vapor movement from the mesophyll evaporative surfaces to the upper air mass and the resistances to this movement

veloping specific cultural practices to reduce water use. Resistances to ET were quantitatively assessed for a St. Augustinegrass turf grown under nonlimiting water conditions (8). This investigation revealed that the external resistances $(R_a + R_c)$ were two to four times greater than the leaf resistance (R_s). The actual values obtained for surface resistance of a St. Augustinegrass turf were comparable to those reported for alfalfa, barley, and sugarbeets. Thus, it was concluded that under well-watered conditions, stomata exhibited limited control of ET from St. Augustinegrass turfs. In other words, the ET rate is controlled to a large extent by factors external to the plant rather than by internal anatomical and physiological factors. These external factors include the number and position of leaves and stems within the turfgrass canopy. A turf with high leaf and stem densities, plus substantial horizontal leaf orientation, can cause greater impairment of the normal upward movement of water vapor and at the same time reduce turbulent eddy movements with a resultant increase in vapor density.

These data indicate a greater potential exists in enhancing resistance to ET by manipulating the canopy structure and density than by altering leaf resistances. The tendency in the past was to emphasize stomatal resistance more than canopy resistance. This, however, does not seem to be the appropriate avenue to pursue for turfgrasses. This premise must be confirmed by studying the ET resistance components across a broad range of turfgrass species.

Controls of evapotranspiration

Three interdependent influences control ET: (a) net radiant energy available at the evaporative surface, (b) those components in the total resistance to water vapor movement from the leaf to the atmosphere, and (c) those factors that influence the vapor pressure gradient between the water at the evaporating surface and the bulk air above. The third dimension will be addressed in this section.

The primary source of water that supplies the evaporative surface is the soil. The soil-water content has a dominant influence on the ET rate. (Soil-water turfgrass relationships are discussed elsewhere and do not require additional attention here.) The only point to be made is that when the leaf-water content decreases as a result of an inadequate soil-water supply to the evaporation sites, the plant may adjust in terms of increased internal diffusive resistance.

Another aspect related to water source is that a functioning root system must be capable of absorbing water at sufficient rates to meet the demands at the evaporative surfaces. Environmental factors that can cause reduced rooting include: excessively high soil temperatures, acidic soils with a pH below 5.0, lack of soil oxygen caused by compaction or waterlogged conditions, and the presence of toxic salts or chemicals, such as certain pesticides (1). Rooting also can be restricted because of certain types of cultural practices, including mowing at too low a cutting height, excessive nitrogen nutritional levels, potassium deficiency, excessive irrigation that causes soil waterlogging, or lack of thatch control (1). (The details of root enhancement from the environmental and cultural standpoints are detailed elsewhere.)

Other factors influencing the vapor pressure gradient between the water source and the evaporating surface involve the environmental variables affecting evaporative demand. Foremost is light, which stimulates opening of the stomates, and as a result, causes most ET to occur during daylight. The typical diurnal variation in net radiation results in a similar diurnal variation in ET rates. Another factor is the water vapor content of the atmosphere. A shift in the vapor content of the air mass above the canopy directly affects the vapor pressure gradient between the stomatal cavity and that air mass. Similarly, wind velocity can affect accumulation of water vapor immediately above and in the upper portion of the turfgrass canopy, which has a direct effect on the vapor pressure gradient. Thus, the ET rate is increased with increases in temperature, decreases in atmospheric water-vapor content, and increasing wind speed, assuming there is an adequate supply of moisture provided to the evaporative sites in the stomatal cavity.

These environmental controls influencing the vapor pressure gradient between evaporative sites in the turfgrass canopy and the above atmosphere vary greatly across the United States and in turn affect ET rates. This variability in evaporative demand will be discussed next.

Interspecies evapotranspiration characterizations

Only six investigations have been published concerning specific comparative characterizations of evapotranspiration rates among perennial grasses maintained under typical turfgrass conditions (3, 4, 10, 11, 19, 20). Most have emphasized warm-season turfgrass species. In nine other studies the specific ET rate of a single grass species is reported (5, 7, 8, 13, 14, 15, 16, 17, 23).

Only three studies have involved detailed comparative characterizations of multiple turfgrass species under controlled cultural conditions where absolute ET values were monitored. Studies in Texas (10) and in Arizona (11) were conducted under nonlimiting water conditions, while an experiment in Israel (3) did not specify the specific soil-water status. The Texas and Arizona studies were conducted in the field with lysimeters positioned in the soil; the Israel study was conducted in the field with containers positioned well above the soil surface. In Texas and Arizona, the ET rates of turfgrasses were less than pan evaporation; in Israel, ET rates of turfgrasses were above pan evaporation.

Results of these studies have been assembled in five tables, followed by a summary table. The data selected are representative of the midsummer growing season under the specific climatic conditions typical of each region. All studies were conducted in the field under mowed turfgrass conditions, except four (8, 16, 17, 19). The duration of the study, cutting height, nitrogen nutritional level, and soil water status are listed, if the information was given in the paper.

St. Augustinegrasses. The ET rates of St. Augustinegrass reported in five different studies are summarized in Table 2. Midsummer ET rates in the 7.7 to 9.6 mm day^{-1} range were reported in three field studies conducted in warm semiarid to arid envi-

TABLE 2 Summary of reported evapotranspiration rates for St. Augustinegrass (*Stenotaphrum* spp.)

Turfgrass species	Location	Evapotranspiration summer mean (mm day^{-1})	Monitoring period (weeks)	Cutting height (cm)	N* rate	Soil water status	Ref[†]
Stenotaphrum secundatum (Texas common)	Texas[‡]	12.2	4	3.8	NL[§]	NL[§]	8
Stenotaphrum secundatum	Arizona	9.6	4	3.8	0.25	NL[§]	11
Stenotaphrum secundatum (Common)	Israel	8.4	12	3	NL[§]	?**	3
Stenotaphrum secundatum (Dwarf)	Israel	7.7	12	3	NL[§]	?**	3
Stenotaphrum secundatum (Texas common)	Texas	6.4	8	5	0.5	NL[§]	7
Stenotaphrum secondatum (Texas common)	Texas	6.3	4	3.8	0.25	NL[§]	10

*kg N are^{-1} month.$^{-1}$
†References presented at end of chapter.
‡Conducted in controlled-climate chamber; no ‡ indicates a field test.
§Nonlimiting.
** Either not stated or not monitored adequately.

TABLE 3 Summary of reported evapotranspiration rates for bermudagrass (*Cynodon* spp.)

Turfgrass species	Location	Evapotranspiration summer mean (mm day^{-1})	Monitoring period (weeks)	Cutting height (cm)	N* rate	Soil water status	Ref[†]
Cynodon hybrid Santa Ana	Arizona	8.7	4	2	0.5	NL[§]	11
Cynodon hybrid Santa Ana	Israel	7.6	12	3	NL[§]	?**	3
Cynodon hybrid Tifgreen	Arizona	8.4	4	2	0.5	NL[§]	11
Cynodon hybrid Tifgreen	Texas	5.4	4	3.8	0.25	NL[§]	10
Cynodon hybrid Tifgreen	Nevada	5.2	8	4	?**	?**	20
Cynodon hybrid Tifway	Texas[‡]	6.8	2	2.5	NL[§]	NL[§]	19
Cynodon hybrid Tifway	Texas	5.9	4	3.8	0.25	NL[§]	10
Cynodon hybrid Tifway	Colorado	4.7	10	2	0.5	NL[§]	4
Cynodon dactylon Common	Arizona	7.2	4	2	0.5	NL[§]	11
Cynodon dactylon Common	Texas	5.8	4	3.8	0.25	NL[§]	10
Cynodon dactylon Common	North Carolina	4.0	16	5	NL[§]	?**	21

*kg N are^{-1} month.$^{-1}$
†References presented at end of chapter.
‡Conducted in controlled-climate chamber; no ‡ indicates a field test.
§Nonlimiting.
** Either not stated or not monitored adequately.

TABLE 4 Summary of reported evapotranspiration rates for zoysiagrass (*Zoysia* spp.)

Turfgrass species	Location	Evapotranspiration summer mean (mm day^{-1})	Monitoring period (weeks)	Cutting height (cm)	N* rate	Soil water status	Ref†
Zoysia japonica Meyer	Texas	7.2	4	5	0.25	NL‡	10
Zoysia japonica Meyer	Texas	5.8	4	3.8	0.25	NL‡	10
Zoysia japonica Local strain	Arizona	7.6	4	2	0.5	NL‡	11
Zoysia matrella	Israel	7.6	12	3	NL‡	?§	3
Zoysia hybrid Emerald	Israel	7.5	12	3	NL‡	?§	3
Zoysia hybrid Emerald	Texas	6.8	4	5	0.25	NL‡	10
Zoysia hybrid Emerald	Texas	4.8	4	3.8	0.25	NL‡	10

*kg N are^{-1} month^{-1}
†References presented at end of chapter.
‡Nonlimiting.
§ Either not stated or not monitored adequately.

ronments. Two studies were conducted under humid field conditions at a lower nitrogen fertility level more typical of home lawn conditions. In this case ET rates of 6.3 to 6.4 mm day^{-1} were observed. The very high rate of 12.2 mm day^{-1} was achieved in the controlled-environment chamber study, where a very high evaporative demand was simulated. St. Augustinegrass is best adapted to warm, humid climates, such as along the Gulf Coast and in Florida (1). Although St. Augustinegrass may have one of the highest potential ET rates under arid conditions of high evaporative demand, these very high levels probably do not occur as frequently in the Gulf Coast region where high atmospheric humidities are common.

Mechanistic studies conducted with St. Augustinegrass (8) demonstrated that a major portion of the resistance to ET is by canopy resistance, the dimensions of which include shoot density, degree of horizontal leaf blade orientation in the canopy, and leaf area. The high ET rates presented in Table 2 for St. Augustinegrass can be interpreted in relation to low canopy resistance plus a rapid vertical leaf extension rate. This concept was confirmed when common and dwarf-type selections were compared (3). The latter had a slower vertical leaf extension rate, a reduced leaf area, and possibly a higher shoot density. It was found to also have a significantly lower ET rate.

A survey of St. Augustinegrass breeder nurseries suggests sufficient genetic diversity exists in canopy structure and vertical leaf extension rate within the *Stenotaphrum* species to develop, via a breeding program, cultivars with a reduced ET rate. However, this needs to be confirmed through a full characterization of canopy resistance and ET rates among the commercially available and near-release St. Augustinegrasses.

Bermudagrasses. Comparative ET rates of turf-type bermudagrasses from eight different studies are summarized in Table 3. Water-use rates varied from 4.0 to 8.7 mm day^{-1}. A broad range in canopy structure, shoot density, leaf area, and vertical leaf extension rate exists among the commercially available bermudagrass cultivars. In the only two comparative studies of *Cynodon* cultivars (10, 11), Santa Ana, Tifway, Tifgreen, and Common varied significantly in ET rates when maintained under a uniform cutting height and nitrogen fertility regime plus a nonlimiting soil water status.

Those bermudagrass studies reporting high ET rates were conducted in climates characterized by a higher evaporative demand. The two lowest ET rates reported were from the more northerly portions of the adaptation range for bermudagrass, specifically Colorado with 4.7 mm day^{-1} and North Carolina with 4.0 mm day^{-1}. In comparisons among cultivars, higher nitrogen levels tended to have a stronger effect on increasing the ET rate than a higher cutting height. This response is probably associated with the enhanced vertical leaf extension rate of these cultivars that have a high nitrogen requirement. This response would reportedly have a significant effect on ET (9).

Zoysiagrass. Only three experiments have been conducted concerning ET rates of zoysiagrasses (Table 4). The zoysiagrasses exhibited ET rates varying from 4.8 to 7.6 mm day^{-1}. Both the number of cultivars and the canopy variation among the commercially available zoysiagrass cultivars are limited. The potential for developing cultivars with lower ET rates is promising and needs assessment. This would involve collecting a broader array of zoysiagrasses to determine how much canopy characteristics contribute to increased ET resistance.

Other warm-season turfgrasses. Data concerning the ET rates of kikuyugrass, centipedegrass, seashore paspalum, bahiagrass, blue grama, and buffalograss are summarized in Table 5. Three of the species, centipedegrass, blue grama, and buffalograss, ranked medium-low in ET rate when maintained at a low nitrogen nutritional level of 0.25 kg N are^{-1} month^{-1}; the two *Pas-*

TABLE 5 Summary of reported evapotranspiration rates of six warm-season perennial turfgrass species

Turfgrass species	Location	Evapotranspiration summer mean (mm day^{-1})	Monitoring period (weeks)	Cutting height (cm)	N* rate	Soil water status	Ref†
Pennisetum clandestinum	Israel	9.0	12	3	NL‡	?§	3
Pennisetum clandestinum	Israel	5.8	12	?	0.5	?§	3
Eremochloa ophiuroides	Israel	8.5	12	3	NL‡	?§	3
Eremochloa ophiuroides Common	Texas	5.5	4	3.8	0.25	NL‡	10
Paspalum vaginatum	Israel	8.1	12	3	NL‡	?§	3
Paspalum vaginatum Adalayd	Texas	6.2	4	3.8	0.25	NL‡	10
Paspalum notatum Argentine	Texas	6.2	4	3.8	0.25	NL‡	10
Bouteloua gracilis Common	Texas	5.7	4	3.8	0.25	NL‡	10
Buchloe dactyloides Common	Texas	5.3	4	3.8	0.25	NL‡	10
Buchloe dactyloides Common	Texas	7.3	4	3.8	0	NL‡	10

*kg N are^{-1} month.$^{-1}$
†References presented at end of chapter.
‡Nonlimiting.
§Either not stated or not monitored adequately.

palum species ranked medium. The potential for centipedegrass to achieve a relatively low ET rate, as reported under Texas conditions, is interesting, especially since centipedegrass has been traditionally grown primarily in humid regions with high rainfall.

Kikuyugrass and bahiagrass possessed high potential ET rates under nonlimiting soil water conditions, possibly because of a more rapid vertical leaf extension rate and high leaf area, plus a reduced horizontal leaf orientation, particularly in the case of bahiagrass.

Two grasses native to the semiarid warm climatic region of the United States, buffalograss and blue grama, exhibited medium-low ET rates when fertilized with nitrogen. In contrast to all other species, buffalograss exhibited a higher ET rate when nitrogen fertilization was decreased (10). This could be associated with improved shoot density due to the preference of buffalograss to soils with minimal nitrogen levels.

Scientifically based information concerning the ET rates of these six warm-season perennial turfgrass species is extremely limited. The extent of their use in the United States also is limited compared with bermudagrass, St. Augustinegrass, and zoysiagrass. Some of these species apparently have the potential to function at inherently low ET rates. Thus, considerably more research is justified on these species under both nonlimiting-water and water-stressed conditions as well as under low nitrogen nutritional levels.

Cool-season turfgrasses. Data concerning ET rates of cool-season turfgrass species are summarized in Table 6. Investigations of cool-season perennial turfgrasses are even more limited than for warm-season turfgrass species. In addition, experiments with two of the four species reported in Table 6 were conducted in warm climatic regions that are on the marginal southernmost limits of adaptation for tall fescue and perennial ryegrass.

Tall fescue proved to have relatively high ET rates when assessed under field conditions. These tests were conducted in climatic regions located on the southernmost limits of adaptation of this species. All data were from locations with extended growing seasons and high evaporative demand, although the atmospheric water-vapor level would be somewhat higher at the Texas test site. It is anticipated that tall fescue would have a somewhat lower mean summer ET rate, if grown and tested in the southern cool humid or transition zones. Also, all tests have been conducted on forage-type tall fescues that have a rapid vertical leaf extension rate, a low shoot density, a wide leaf, and a fairly erect leaf orientation. It will be interesting to compare these ET results with rates for the newer turf-type tall fescue cultivars.

There are only two reports of ET rates for perennial ryegrasses, with the range between the two sites being great: 6.6 versus 11.2 mm day^{-1}. Both sites were in the southernmost limits of adaptation in semiarid climates characterized by high evaporative demand. The perennial ryegrasses are best adapted to the cool humid and transition climatic zones. Much research is needed concerning comparative ET rates among the turf-type perennial ryegrass cultivars grown in their normal region of climatic adaptation.

Two of the three studies concerning ET rates of creeping bentgrass were conducted on Penncross in controlled-environ-

TABLE 6 Summary of reported evapotranspiration rates for four perennial cool-season turfgrass species

Turfgrass species	Location	Evapotranspiration summer mean (mm day^{-1})	Monitoring period (weeks)	Cutting height (cm)	N* rate	Soil water status	Ref†
Festuca arundinacea Alta	Israel	12.6	12	3	NL§	?**	3
Festuca arundinacea Kentucky 31	Arizona	10.6	4	4	0.25	NL§	11
Festuca arundinacea Kentucky 31	Texas	7.2	4	3.8	0.25	NL§	10
Lolium perenne Pennfine	Israel	11.2	12	3	NL§	?**	3
Lolium perenne	California	6.6	8	?	?**	?**	15
Agrostis palustris Penncross	Texas‡	9.7	2	2.5	NL§	NL§	19
Agrostis palustris Penncross	Michigan	7.9	1	2.5	NL§	NL§	17
Agrostis palustris Penncross	Arizona	5.0	4	0.6	0.5	NL§	12
Poa pratensis Baron	Kansas	4.1	1.5	6.3	0.25	PWS‡‡	14
Poa pratensis Baron	Nebraska‡	4.3	1	5	NL§	NL§	16
Poa pratensis Merion	Nebraska‡	6.6	1	5	NL§	NL§	16
Poa pratensis Merion	Colorado	4.9	8	2	0.5	NL§	5
Poa pratensis plus *Festuca rubra* polystand	Nevada	5.2	16	4	?**	?**	20

*kg N are^{-1} month.$^{-1}$
†References presented at end of chapter.
‡Conducted in controlled-climate chamber; no ‡ indicates a field test.
§Nonlimiting.
**Either not stated or not monitored adequately.
††Progressive water stress.

ment growth chambers with a high evaporative demand and both nonlimiting nutrient level and soil moisture status. The data reported (7.9 and 9.7 mm day^{-1}) should approach the potential ET, with ET levels in the field projected to be lower as supported by the third study where the cutting height was much lower. Penncross is a vigorous species with a fairly rapid vertical leaf extension rate. Thus, it is very possible that other cultivars of creeping bentgrass, as well as some of the colonial bentgrasses, might possess lower ET rates. More research needs to be conducted concerning ET levels of bentgrasses under both close-cut greens and higher-mowed cultural regimes.

There are four reports of ET rates for Kentucky bluegrass, two with Merion and two with Baron. The ET rates for this species were ranked surprisingly low, with Baron ranking slightly lower than Merion. Two of the experiments were conducted in controlled-environment growth chambers, simulating a high evaporative demand as well as both nonlimiting nutrient level and soil water status. The resultant ET rates were 6.6 mm day^{-1} for Merion and 4.3 mm day^{-1} for Baron.

One study involved a polystand of Kentucky bluegrass and fine-leafed fescues in Reno, Nevada, a semiarid region of high evaporative demand. This polystand, composed of the two major cool-season turfgrass species, ranked medium-low in ET rate.

This summary of ET rates across the cool-season turfgrasses emphasizes that the amount of information available is deficient, especially in comparison with the warm-season turfgrasses. In addition, most field experiments on cool-season grasses have been conducted in climatic regions representing the marginal southernmost regions of adaptation. Thus, there is a potential for reduced canopy resistance caused by shoot thinning due to heat stress. Research is needed concerning the specific ET rates of cool-season grasses grown in optimum environments representative of where each is commonly grown. Furthermore, there is no information on the ET rates of turf-type tall fescues, the fine-leafed fescues (including red, chewings, sheep, and hard), meadow fescue, colonial bentgrass, and a broader range of turf-type perennial ryegrasses.

The more commonly used polystands of cool-season grass species also need to be characterized, since these turfgrass communities are widely used in cool climatic regions, whereas most warm-season turfgrasses are commonly grown as monostands. There is the possibility that two or three selected species could be combined to form a more horizontal composite leaf orientation characterized by increased canopy resistance.

Interspecies overview. Current knowledge of the mean summer ET rates of perennial grasses maintained under typical turf conditions in the field, is summarized in Table 7. ET rates are ex-

TABLE 7 Range in reported summer mean evapotranspiration rates summarized by turfgrass species

Scientific name	Common name	Evapotranspiration summer mean field conditions	
		(mm day^{-1})	(in/wk)
Festuca arundinacea	Tall fescue	7.2–12.6	2.0–3.5
Lolium perenne	Perennial ryegrass	6.6–11.2	1.8–3.1
Stenotaphrum secundatum	St. Augustinegrass	6.3– 9.6	1.7–2.6
Paspalum vaginatum	Seashore paspalum	6.2– 8.1	1.7–2.2
Paspalum notatum	Bahiagrass	6.2	1.7
Pennisetum clandestinum	Kikuyugrass	5.8– 9.0	1.6–2.5
Agrostis palustris	Creeping bentgrass	5.0– 9.7	1.3–2.7
Bouteloua gracilis	Blue grama	5.7	1.6
Eremochloa orphiuroides	Centipedegrass	5.5– 8.5	1.5–2.3
Buchloe dactyloides	Buffalograss	5.3– 7.3	1.5–2.0
Cynodon spp.	Bermudagrass	4.0– 8.7	1.0–2.2
Zoysia spp.	Zoysiagrass	4.8– 7.6	1.3–2.1
Poa pratensis	Kentucky bluegrass	4.1– 6.6	1.1–1.8

pressed in inches per week as well as in millimeters per day.

At the time this investigator initiated comparative studies of potential ET among species and among cultivars within species several knowledgeable agronomists predicted that no significant differences would be found. A study of the data reveals significant differences in water use between species and within species of perennial turfgrasses. Some crop scientists have reported no differences in water-use efficiency at the intraspecies level; other scientists found differences among forage species. In the case of turfgrasses, we're concerned with the less complex process of ET. The relationships become much more complex genetically where plant/water relations are related not only to vegetative growth but also to yield and associated metabolic processes. Fortunately, we are not faced with this more complex problem in our quest for turfgrass water conservation.

Several striking aspects are evident in Table 7. First, information concerning more than 50 percent of the cool-season turfgrass species is lacking. Among the turfgrass species ranking low in ET rates is Kentucky bluegrass, which may surprise many. In addition, certain *Zoysia* and *Cynodon* species apparently have good potential as low water-use species. Another interesting species, centipedegrass, had a medium-low ET rate under the more humid test conditions of Texas. It would be interesting to know the specific ET rates of such fine-leafed fescues as red, chewings, sheep, and hard. These species may also rank low in ET rates.

From the mechanistic standpoint all four turfgrass species just discussed tend to have a combination of two or more of the following characteristics: a slower vertical leaf extension rate, lower leaf area, higher shoot density, or more horizontal leaf orientation, compared with those perennial turfgrasses that possess higher ET rates.

It may seem surprising that the only native turfgrass species listed, buffalograss and blue grama, did not rank lowest in ET rates. This, however, is not an uncommon response for plants adapted to semiarid/arid climates. For example, many dicotyledonous species adapted to semiarid and arid climatic zones typically have high ET rates. This emphasizes the fact that low water use is distinctly different from drought resistance.

Four warm-season species had relatively high ET rates: St. Augustinegrass, seashore paspalum, bahiagrass, and kikuyugrass. These high ET rates can be readily interpreted in relation to canopy resistance. All three species tend to have medium-to-low shoot density, a more erect leaf orientation, and a wider leaf, particularly St. Augustinegrass and bahiagrass, as well as a very rapid vertical leaf extension rate.

There is a developing tendency to classify C-3 cool-season turfgrasses as possessing higher ET rates than the C-4 warm-season turfgrasses (3, 4, 11). However, the few cool-season turfgrasses compared to date were evaluated at locations well out of their normal range in adaptation. In addition, ET rates of many cool-season turfgrass species, such as the fine-leafed fescues, colonial bentgrass, and meadow fescue, have not been assessed. Thus, it would be premature to classify C-3 and C-4 turfgrasses into high and low ET groupings. More comprehensive investigations might prove that the range in ET rates of the C-3 cool-season and the C-4 warm-season perennial grasses may vary similarly.

The ET rate of a turf can be altered substantially by any change in cultural practices or environmental conditions that could cause a change in canopy density, leaf area, leaf orientation, or vertical leaf extension rate. Thus, the summer-mean ET rates of turfgrasses can be strongly affected by their ability to adapt to a particular climatic region. The intensity of the evaporative demand of a region, the occurrence of environmental or pest stresses, and the cutting height and nitrogen nutritional level selected can cause substantial shifts in the ET rate and the resultant comparative rankings among species.

The great variability in existing data summarized in Table 8 emphasizes the importance of conducting comparative ET rate studies at the interspecies level in a controlled, stress-free environment. Such studies are best conducted in controlled-environment growth chambers with light intensities, light quality, carbon dioxide levels, and wind velocities representative of field conditions. A pretest acclimatization period should be selected to allow optimum shoot and root growth of the species being assessed. The actual ET monitoring test should be conducted initially under nonlimiting water conditions to assess the potential ET maximums. Subsequently, experiments should be conducted in a controlled environment in which progressive water stress is imposed and the associated ET rates are carefully monitored. This experimental approach ensures valid absolute comparative assessments that can be interpreted in relation to the diverse climatic regions of the United States.

Intraspecies evapotranspiration characterizations

Little published research exists regarding comparative ET rates among cultivars within a turfgrass species (2, 3, 10, 11, 16). The only two research reports available that assessed intraspecies ET rates among a large number of cultivars were made in 1973 in Michigan (2) and in 1978 in Nebraska (16). Both studies assessed the comparative ET rates among Kentucky bluegrass (*Poa pratensis* L.) cultivars in controlled-environment growth chambers. The environmental regime imposed a high evaporative demand, and the root zone was manipulated to achieve a nonlimiting soil-water status. The ET-rate test period was brief in replicated segments, following a 3- to 4-week acclimatization period in a controlled-environment growth chamber.

Comparative rankings between these two studies in terms of ET rates among Kentucky bluegrass cultivars are summarized in Table 8. Significant differences in ET rates were evident in both studies. Of seven cultivars common to both experiments, Sydsport, Fylking, and Bensun ranked comparable in ET; Newport, Baron, Merion, Nugget, and Park varied substantially in ET. Furthermore, Shearman and Riordan (private communication, 1983) have observed significant seasonal variations in relative ET rankings among Kentucky bluegrass cultivars. These differentials in ET within individual cultivars may be due to seasonal climatic variations under which the turfs were originally propagated in the field. This may have occurred even though there was an acclimatization period in a growth chamber before the ET measurements were initiated.

Shifts in comparative ET rankings at the intraspecies level are not unexpected, if assessed from a mechanistic standpoint. The East Lansing, Michigan site represents more optimum temperature, moisture, and soil conditions for growth of Kentucky bluegrass during the summer; substantial summer heat stress plus periodic drought stress did occur at the Lincoln, Nebraska site. Relative climatic adaptation of cultivars, as it influences root and shoot growth dimensions and associated resistances to ET, may singly or in combination cause these shifts.

First, the ability to absorb available soil moisture via the root system can substantially affect the ET rate. Turfgrasses growing in areas characterized by heat stress, waterlogged soils, soil compaction, salinity, or extremes in acidic or alkaline soil reaction may have significantly impaired root systems (1). Cultural practices, such as very close mowing, excessively high nitrogen rates, excessive irrigation, thatch accumulation, and improper usage of certain pesticides, may also restrict root growth (1). Turfgrass cultivars that vary in rooting capability under en-

TABLE 8 Comparison of relative evapotranspiration rates among Kentucky bluegrass (*Poa pratensis*) cultivars when grown in two different climatic regions

Relative cultivar evapotranspiration ranking	Kentucky bluegrass cultivars	
	Michigan	Nebraska
Low	Prato Cougar Delta Kenblue	Enoble Adelphi A-20 Newport Baron
Medium-low	Pennstar Park Nugget Windsor	Cheri Touchdown
Medium	Merion Galaxy Monopoly Baron	Parade Bensun Victa
Medium-high	Bensun Newport Fylking	Park Fylking South Dakota
High	Sodco Sydsport	Bristol Bonnieblue Nugget Majestic Birka Sydsport Merion

SOURCE: Beard, Eaton, and Yoder, 1973; Shearman, 1978.

vironmental stresses and adverse cultural circumstances can be expected to possess corresponding differentials in ET rates.

Environmental factors that cause intraspecies variations in shoot growth that affect canopy resistance to evapotranspiration could contribute to corresponding shifts in ET rate. For example, such stresses as heat, cold, water, shade, and wear are known to alter shoot density, leaf area, vertical leaf extension rate, stomatal density, and relative horizontal leaf orientation (1). Thus, a decreased ET rate could be expected in climatic regions where certain turfgrass cultivars are (a) best adapted and (b) least prone to environmental and biological stresses that negatively affect shoot density and canopy structure. This is due to an increase in canopy resistance to evapotranspiration.

The great variability in the existing data summarized here emphasizes the need to conduct comparative ET rate studies at the intraspecies level in a controlled stress-free environment to obtain valid assessments of the comparative maximum ET potential. Concurrently, detailed basic research is needed concerning the specific mechanisms controlling ET in turfgrasses as well as their relative significance. Turfgrass researchers located in different climatic regions could assess the specific adaptation of the individual cultivars to their particular environments and soils, and this information could be combined with an understanding of the basic concepts and factors controlling resistances to ET for that species. As a result, valid conclusions can be drawn regarding comparative ET rates and the cultivars that would be preferred for use in each individual climatic/soil region from a water conservation standpoint.

It has been established that canopy resistance has a substantial role in controlling ET from turfs and that substantial differences in shoot density, leaf area, horizontal leaf orientation, and vertical leaf extension rate exist at the intraspecies level for several turfgrass species. Thus, there may be as much variation in the ET rate at the intraspecies level, for a number of species, as at the interspecies level. Additional research is needed to confirm this. Nevertheless, sufficient basic information is available to conclude that specific cultivars can be selected that would have lower ET rates. This emphasizes the need to characterize the commercially available cultivars across the full range of perennial turfgrass species. This information will aid the general public and the turf professional in selecting cultivars that will contribute to water conservation strategies. It is also an important initial descriptive phase leading to breeding water-conserving turfgrasses.

Plant improvement—breeding

In the past no effort was made to breed and select turfgrass cultivars possessing reduced water-use rates. It is hoped that this situation will change. Recent investigations have resulted in new concepts regarding the relative significance of canopy resistance to ET, including its greater importance compared with stomatal resistance (6, 8, 9). While these investigations have led the way, there is still a need to characterize the resistances to ET across a broader range of commonly used turfgrasses. Such research will provide the plant breeder with specific guidelines for predicting which turfgrasses have a reduced water-use rate. Equally significant, it appears that several key turfgrass characteristics can be assessed rapidly, thereby facilitating assessment of large plant populations. The net result can be increased probabilities for selecting turfgrasses with reduced ET rates.

The stomatal resistance component also must be considered, even though it is less important than canopy resistance. Only one investigation has compared stomatal densities at the intraspecies level. Significant differentials were reported within the bentgrasses (*Agrostis* spp.) (17). Obviously, additional research is needed concerning the genetic diversity in stomatal resistance.

The phase before implementation by the grass breeder involves assessments across a broad range in plant types within a species which influence canopy resistance, followed by correlations of these plant types with the measured ET rates. This approach would confirm the validity of these concepts for use in breeding. Studies are needed for each major turfgrass species.

The next phase is to implement the breeding program. This involves collecting plant material possessing components of canopy resistance that will reduce the ET rate. This plant material would then be incorporated into the breeding program. These experiments should include inheritance studies of these key characteristics within each grass species. Combining these approaches with established breeding techniques should result in improved grasses from a water conservation standpoint.

The essence of this discussion is that it is feasible to breed and select turfgrass cultivars with reduced evapotranspiration rates, thereby contributing to water conservation. Based on our current understanding of the plant components of canopy resistance to ET, it is evident that substantial genetic diversity exists within a number of turfgrass species to allow breeders to develop plant selections possessing increased canopy resistance.

Research needs

Fundamental investigations into resistance mechanisms and comparative evapotranspiration rates of cool-season turfgrasses, at both the interspecies and intraspecies levels, are in their infancy. Our basic pool of knowledge is extremely small. Nevertheless, substantial advances must be made during the 1980s to meet the challenges of a decreasing water supply while maintaining the desired turfgrass quality under even more intense use, especially in urban areas. Much of this research will require fairly sophisticated techniques and research facilities. This means that funds for research focusing on water conservation must be greater than for any other previous major advance in turfgrass science.

A flow plan for the orderly coordinated pursuit of research concerning water conservation of turfgrasses is shown in figure 4. The first step in any research program is the descriptive phase; in this case, it involves characterization of comparative ET rates at the interspecies level. Investigation should first be conducted under nonlimiting moisture conditions, followed by a second series of comparisons conducted under progressive water stress. In one series of experiments, relative rankings of warm-season turfgrasses under nonlimiting and under progressive moisture stress were assessed (10). With the exception of bahiagrass,

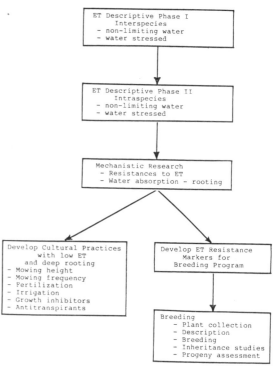

FIGURE 4. A flow plan for the coordinated pursuit of research concerning the development of water conserving turfgrass and cultural systems.

there were no large shifts in comparative water-use rates between the high and low ET ranges. Nevertheless, additional research must be conducted across the broad range of turfgrass species to evaluate fully the comparative effects of plant water stress on ET rates.

Following completion of the interspecies comparisons, the second phase would involve ET rate comparisons at the intraspecies level across the commercially available cultivars. Results can provide information of immediate value to practitioners in the field. Equally important, the descriptive characterization will indicate the range in genetic diversity available in terms of low ET rates. These data will facilitate selection of plant materials for use in studying the mechanisms of resistance to evapotranspiration, the third research phase. Basic research techniques for this work have already been established (9). Subsequent followup has demonstrated that the same techniques can be used in field plot research. The relative importance of various resistances to ET has been established for St. Augustinegrass; so have the key plant components contributing to canopy resistance (6). Similar studies need to be conducted across a broader range of turfgrass species.

The basic concepts evolving from this research can be utilized in two ways. One thrust will provide a sound basis for developing cultural strategies in water conservation. An example illustrating the importance of basic studies, from which practical field techniques can evolve, is the development of growth inhibitors for water conservation. Johns and Beard (7) first reported that growth inhibitors could significantly reduce ET rates of turfgrasses. In addition, these mechanistic studies have provided a sound basis for interpreting how such cultural practices as a lower mowing height or a reduced nitrogen nutritional level cause a decreased ET rate. The second thrust involves identification of specific plant markers that can be used to breed turfgrasses with low ET rates. This aspect of water conservation research needs to be initiated as soon as the mechanistic studies are completed on each major turfgrass species.

Selection of the proper techniques for ET experiments are critical to avoid confounding by environmental variables and cultural practices. The preferred approach to achieve valid comparative information is to conduct these studies in controlled stress-free environments. Environmental simulation chambers have been developed and successfully tested for such experiments (6, 8, 9). These controlled-environment experiments should be confirmed through field experiments under nonlimiting water conditions.

The experiments should be designed to separate shoot resistances to ET from resistances related to water absorption by the root system. The environmental and cultural conditions selected should not place adverse stresses on the root system. Of equal importance is the use of fritted clay as the root-zone media in mini-lysimeter experiments (24). Fritted clay has an average dry bulk density of 0.67 ton·m^{-3}, particle density of 2.50 tons·m^{-3}, and total porosity of 0.73. After drainage, the material holds 0.31 by volume of plant-available water, and has an air-filled porosity of 0.28. It retains a large quantity of plant extractable water over a broad range of pressure potentials.

It is critical that the controlled-environment conditions under which ET rates are monitored represent normal field conditions. This includes a high light intensity possessing a reasonable spectral distribution in the photosynthetically active range.

A minimum PAR of 400 watts m^{-2} is needed. A carbon dioxide monitoring/injection system is also required to ensure that the carbon dioxide levels are in the normal field range. It is particularly important that the CO_2 be carefully monitored in closed systems, especially when C-4 grasses are utilized. Finally, a minimum wind velocity of 1.0 meter per second should be maintained across the grass surface to minimize water vapor boundary layer effects.

In the case of field experiments mini-lysimeters are preferred, along with the fritted-clay root zone. Adequate under drainage is critical to provide rapid downward movement of gravitational water and to avoid upward capillary movement of water from underlying zones. Finally, it is important to select a cultural regime, including cutting height and nitrogen nutritional level, that is most appropriate to the objectives of the research. In some studies it might be preferable to hold these cultural factors constant. However, it is important to monitor ultimately the ET rates of individual turfgrass species and cultivars under their respective optimum cultural regimes.

Note: It is hoped that the reader will recognize that this author has prepared this chapter while faced with substantial deficiencies in sound basic research information. Thus, there are proposed hypotheses and projected interpretations of mechanistic dimensions affecting some reported ET rates, as well as projected ET rates for turfgrass species that have not been studied. This is a somewhat tenuous position, but it was undertaken with the expectation that it will stimulate research efforts and water conservation strategies for turfgrasses. When an author writes a similar chapter in another decade, let us hope that there will be much more research information available.

Acknowledgments

The author wishes to acknowledge the excellent research of four former graduate students whose data have contributed substantially toward preparation of this chapter. They are Dr. Don Johns, Ki Sun Kim, Gwen K. Stahnke, and Dr. Robert C. Shearman. In addition, the critiques of fellow colleagues, Dr. C. H. M. van Bavel and Dr. Charles W. Wendt, are appreciated very much as is the assistance of Harriet J. Beard in preparing the manuscript. Equally important is the contribution of the United States Golf Association, as most of the turfgrass water conservation research conducted at Texas A & M University by the Texas Agricultural Experiment Station and summarized in this chapter was partially supported by USGA grants.

Literature cited

1. BEARD, J. B.
    1973. Turfgrass: Science and Culture. Prentice-Hall, Inc., Englewood Cliffs, N.J. 658 pp.
2. BEARD, J. B., W. J. EATON, and R. L. YODER
    1973. Physiology research: chemical growth regulators, water use rates, thatch causes, and low temperature kill. Proc. Michigan Turfgrass Conf. 2:27–33.

3. BIRAN, I., B. BRAVDO, I. BUSHKIN-HARAV, and E. RAWITZ
   1981. Water consumption and growth rate of 11 turfgrasses as affected by mowing height, irrigation frequency, and soil moisture. Agron. J. 75:85–90.
4. DANIELSON, R. E., C. M. FELDHAKE, and W. E. HART
   1981. Urban lawn irrigation and management practices for water saving with minimum effect on lawn quality. Colorado Water Resources Research Institute. Report 106. 120 pp.
5. DANIELSON, R. E., W. E. HART, C. M. FELDHAKE, and P. M. HAW
   1979. Water requirements for urban lawns. Colorado Water Resources Research Institute. 90 pp.
6. JOHNS, D.
   1980. Resistances to evapotranspiration from St. Augustinegrass (*Stenotaphrum secundatum* [Walt.] Kuntze) turf. Ph.D Thesis. Texas A & M Univ., College Station, Tex. 88 pp.
7. JOHNS, D., and J. B. BEARD
   1981. Reducing turfgrass transpiration using a growth inhibitor. Agron. Absts. p. 126.
8. JOHNS, D., J. B. BEARD, and C. H. M. VAN BAVEL
   1983. Resistances to evapotranspiration from a St. Augustinegrass turf canopy. Agron. J. 75:419–422.
9. JOHNS, D., C. H. M. VAN BAVEL, and J. B. BEARD
   1981. Determination of the resistance to sensible heat flux density from turfgrass for estimation of its evapotranspiration rate. Agric. Meteorology 25:15–25.
10. KIM, K. S.
    1983. Comparative evapotranspiration rates of thirteen turfgrasses grown under both nonlimiting soil moisture and progressive water stress conditions. M.S. Thesis. Texas A & M Univ., College Station, Tex. 64 pp.
11. KNEEBONE, W. R., and I. L. PEPPER
    1982. Consumptive water use by sub-irrigated turfgrasses under desert conditions. Agron. J. 74:419–423.
12. KRANS, J. V. and G. V. JOHNSON
    1974. Subirrigation and fertilization of bentgrass during prolonged heat stress. Proc. Second International Turfgrass Research Conf. 2:527–533.
13. MANTELL, A.
    1966. Effect of irrigation frequency and nitrogen fertilization on growth and water use of a kikuyugrass lawn (*Pennisetum clandestinum* Hochst.). Agron. J. 58(6):559–561.
14. O'NEIL, K. J., and R. N. CARROW
    1982. Kentucky bluegrass growth and water use under different soil compaction and irrigation regimes. Agron. J. 74:933–936.
15. PRUITT, W. O.
    1964. Evapotranspiration—A guide to irrigation. California Turfgrass Culture 14(4):27–32.
16. SHEARMAN, R. C.
    1978. Water use rates and associated plant characteristics for twenty Kentucky bluegrass cultivars. Turfgrass Research Summary Progress Report No. 78–1. Department of Horticulture, Univ. of Nebr. p. 82.
17. SHEARMAN, R. C., and J. B. BEARD
    1972. Stomatal density and distribution in *Agrostis* as influenced by species, cultivar, and leaf blade surface and position. Crop Sci. 12:822–823.
18. SLATYER, R. O.
    1967. Plant-Water Relationships. Academic Press, N.Y. 366 pp.
19. STAHNKE, G. K.
    1981. Evaluation of antitranspirants on creeping bentgrass (*Agrostis palutris* Huds., cv. 'Penncross') and bermudagrass (*Cynodon dactylon* [L.] Pers. X *Cynodon transvaalensis* Burtt-Davy, cv. 'Tifway'). M.S. Thesis. Texas A & M Univ., College Station, Tex. 70 pp.
20. TOVEY, R., J. S. SPENCER, and D. C. MUCKEL
    1969. Turfgrass evapotranspiration. Agron. J. 61:863–867.
21. VAN BAVEL, C. H. M.
    1966. Potential evaporation: the combination concept and its experimental verification. Water Resources Research 2(3):455–468.
22. ———
    1967. Changes in canopy resistance to water loss from alfalfa induced by soil water depletion. Agric. Meteorology 4:165–176.
23. VAN BAVEL, C. H. M., and D. G. HARRIS
    1962. Evapotranspiration rates from bermudagrass and corn at Raleigh, N.C. Agron. J. 54:319–322.
24. VAN BAVEL, C. H. M., R. LASCANO, and D. R. WILSON
    1978. Water relations of fritted clay. Soil Science Society of America J. 42:657–659.

Practicum

6 Turfgrass culture and water use

ROBERT C. SHEARMAN

Robert C. Shearman, associate professor-turf, has been affiliated with the Department of Horticulture at the University of Nebraska for 10 years. He has conducted research on turfgrass water-related management problems during this time, and has written numerous scientific and popular articles on subjects relating to turfgrass culture.

Turfgrass managers interested in conserving water should mow high and frequently, fertilize to meet nutritional needs, and irrigate infrequently based on soil-moisture sensing devices rather than on set schedules.

Turfgrasses use water for growth and the process of evapotranspiration. Their consumptive water use increases with:

- Vegetative cover,
- Growth rate,
- Length of growing season, and
- Conditions of high evaporative demand (i.e., high temperature, low humidity, and air turbulence). Factors that stimulate shoot growth and leaf area or reduce canopy resistance result in an increased water-use rate.

Reduced water use does not equate with turfgrass drought tolerance, but it can be an important factor in drought avoidance because a low water-use rate would extend the time that available soil moisture could be used as long as that available moisture was within the root zone. Relating turfgrass water use to water-use efficiency is an important concept.

Turfgrass water-use efficiency should be based on the water required to produce a desired level of turfgrass functional quality rather than on dry matter production as it is in other crops. Therefore, an acceptable turfgrass water-use level would be higher for a golf green than for low-maintenance utility turf.

Turfgrass water use increases with mowing height and declines with frequency of mowing. Furthermore, turfs mowed with dull, improperly adjusted mowers are less efficient in water use than those mowed with sharp mowers.

Turfgrass managers concerned with water conservation should:

- Mow with properly adjusted mowers at higher end of the optimum turfgrass species mowing-height range, and
- Mow as frequently as possible based on economic and physical limitations.

Turfgrass water use and water-use efficiency increase with nitrogen fertilization rate until a critical species level is obtained. At this critical point, water-use efficiency declines because excessive nitrogen fertilization reduces depth and extent of rooting more dramatically than it does shoot growth. Turfgrass managers interested in conserving water should mow high and frequently, and they should supply only enough nitrogen to meet turfgrass nutritional needs and maintain a desired recuperative rate. On cool-season turfgrasses, emphasis should be placed on fall-applied nitrogen rather than on spring applications.

Maintenance of adequate potassium levels reduces turfgrass wilting tendency. Potassium levels should be carefully monitored and kept readily available, particularly in periods of high temperature and drought stress.

Turfs growing under frequent irrigation and high available soil-moisture contents have higher water-use rates than those at low available soil-moisture contents. Therefore, irrigation frequency can be adjusted to reduce water-use requirements of turfs.

Adjusting irrigation scheduling based on moisture-sensing devices and pan evaporation—as opposed to scheduled irrigation—can reduce water consumption by 40 to 50 percent without reducing turfgrass quality. Moisture-sensing devices are particularly effective in reducing water consumption during periods of low evaporative demand.

Syringing during periods of high temperature stress increases turfgrass water-use rate, but it decreases wilting tendency.

Where supplemental irrigation is unavailable, turfgrass managers will have to:

- Select drought-tolerant and resistant turfgrass species and cultivars,
- Employ cultural practices that harden the plant to moisture stress and enhance drought tolerance, or
- Use alternative plant materials and surface covers.

Soil cultivation and wetting agents can be used to enhance infiltration, percolation, and soil-moisture content. Water-use efficiency is enhanced on noncompacted turfgrass soils.

Antitranspirants can reduce water consumption, but they are potentially phytotoxic and likely are not economical for use at this time.

Use of plant-growth regulators in turfgrass management is limited. Plant-growth regulators affect turfgrass water use as does mowing. More research is needed as to their potential influence for reducing turfgrass water use.

Turfgrass managers interested in conserving water should manipulate the primary cultural practices of mowing, fertilizing, and irrigating. Specific amounts of water conservation cannot precisely be stated because differences will be relative to species, cultivar, location, use, and evaporative demand. As more research is accomplished in the area of turfgrass water conservation, drought resistance and drought tolerance, more specific recommendations will be feasible. However, at this time relative comparisons can be made and manipulation of cultural practices can significantly reduce water use and enhance water-use efficiency.

6 Turfgrass culture and water use

Robert C. Shearman

Water requirements of turfgrasses are important in their selection, adaptation, and use, particularly in areas and times when water for turfgrass culture and maintenance is restricted. Turfgrasses require water for growth and evapotranspiration. Turfgrass water use has been defined as the total amount of water used for plant growth plus water lost through transpiration and evaporation from plant and soil surfaces (3). In this regard, turfgrass water use increases with growth rate (4), length of growing season (3, 74), and the presence of turfgrass cover versus bare soil (71). It is reasonable to consider that factors influencing turfgrass leaf, shoot and root growth, or leaf area and extent of rooting would influence water use. Turfgrass researchers and managers should develop a thorough understanding of factors influencing turfgrass water use. This knowledge could be used to improve irrigation systems and watering practices, reduce maintenance costs, and conserve water.

Efficiency of water use has been related to water consumption per unit of dry matter production (5, 16, 36, 64). Water-use efficiency in turf is not as easily defined as it is in crop production, since turfgrass managers are not interested in clipping yields or dry matter production. A more realistic assessment of turfgrass water-use efficiency would relate water consumption to turfgrass quality. This chapter will discuss turfgrass cultural practices and their influence on turfgrass water use and water-use efficiency. Readers should keep in mind that low water use is not directly related to drought resistance in a turfgrass species, but could be important to drought avoidance and maintenance of desired turfgrass quality.

Cultural practices

Mowing, fertilization, and irrigation are the primary cultural practices that affect turfgrass growth and development. They have direct effects on growth rate, leaf surface area, canopy resistance, and extent of rooting, and therefore are expected to influence water use. Primary cultural practices alone or in combination can be manipulated to minimize water loss and enhance water conservation.

Mowing. Turfgrass mowing height, frequency, and equipment directly and indirectly affect turfgrass growth, development, quality, and water use, and scientists (3, 34) have summarized many physiological and morphological responses of turfgrass shoots and roots to variations in mowing height and frequency. Increased water use has been associated with higher canopy heights in both warm- and cool-season turfgrass (13, 19, 35, 40, 56, 64, 69). Water use of 'Penncross' creeping bentgrass (*Agrostis palustris* Huds.) was found to increase as mowing height increased from 0.6 cm (0.25 inch) to 12.5 cm (5.0 inches) (56). Turfs mowed at 12.5 cm used twice as much water as those mowed at 0.6 cm and those mowed at 2.5 cm (1.0 inch) used 56 percent more water than those mowed at 0.6 cm. It was also found that a well-watered common bermudagrass [*Cynodon dactylon* (L.) Pers.] transpired 4.8 mm of water daily at a plant height of 15 cm and 3 mm daily at 2.5 cm (13). Clipping resulted in a 37 percent decline in transpiration. Similar results were demonstrated with perennial ryegrass (*Lolium perenne* L.) mowed at 5 cm and 25 cm (40). Increased water loss beneath a 'Merion' Kentucky bluegrass turf has been related to increased mowing height and a subsequent increased depth and extent of rooting (35). Increased water use associated with mowing height has been seen as a result of added plant surface area exposed to desiccating conditions (56). Studies of responses of 'Seaside' creeping bentgrass and 'Highland' colonial bentgrass (*A. tenuis* Sibth.) to mowing height indicated increased plant density, but reduced verdure and rooting with low mowing (32). Similar effects of mowing treatments on leaf, shoot, and root development have also been reported (17, 18, 45, 52, 72).

Reduced water use found with lower mowing heights would most likely be associated with an increased canopy resistance and a decreased leaf-area index, since low-cut turfs have a high shoot density and a dense, tight canopy. On the other hand, high-cut turfs have an open canopy with reduced shoot density. High-cut turfs would have a low canopy resistance and a high-leaf area index that would enhance evapotranspirational water loss. Water-use efficiency of taller turfs would likely be increased due to enhanced depth and extent of rooting associated with the high cut.

It has been found that evapotranspiration from a St. Augustinegrass [*Stenotaphrum secundatum* (Walt) Kuntze] turf increased substantially with days following mowing, indicating a greater leaf area attributed to growth occurring between mowings. Evaluations of mowings made biweekly, weekly, and six times weekly on 'Penncross' in 1973 indicated a 41 percent increase in water use for turfs mowed biweekly compared with those mowed six times weekly (56). These results were contradicted in a 1980 report (23). Authors of the earlier study offered no explanation for the results obtained in their mowing frequency study. Turfs in their investigation were maintained at a relatively high nitrogen nutrition level [5.0 g N/m^2 (1.0 lb N/1000 ft^2)] and were well watered, which may have contributed to an excessive growth rate, succulent plant tissue and increased water use. Other studies indicated that leaf area, shoot size, and extent of rooting decreased, but shoot density and tissue succulence increased with frequent mowing (9, 22, 30, 31, 32, 35, 49). Further research is needed to clarify the effects of mowing frequency on turfgrass evapotranspiration rates.

It is apparent that within turfgrass species mowing height and frequency tolerance ranges there is room to manipulate these cultural practices to enhance depth and extent of rooting and minimize excessive leaf area. This would reduce evapotranspiration and enhance soil-moisture extraction, thus increasing water-use efficiency. This was demonstrated in research with St. Augustinegrass when excessive leaf area was eliminated and transpiration was reduced with mowing frequency (23). Additional research with C-3 (cool-season) and C-4 (warm-season) pathway turfgrasses, relating mowing height and frequency responses to evapotranspiration rates, is needed. Differences in water consumption between C-3 and C-4 pathway turfgrasses were demonstrated, with C-3 plants using more water, even though both groups had similar growth rates (4). Care should be taken in interpreting these results and those of others (26, 71), since the studies were conducted where C-4 plants were more suitably adapted than C-3 plants.

Mower condition may also play a part in turfgrass water use. It has been speculated that greater water loss could occur from turfs mowed with a dull or improperly adjusted mower as opposed to those mowed with a sharp mower blade (3). The increased water loss could be associated with increased tissue mutilation as a result of dull mower blade injury, but no data exist to support this hypothesis (3). In a study of the effect of dull mowers on a field-grown 'Park' Kentucky bluegrass turf, it was found that water use was reduced with dull mower treatments under field and controlled environment conditions (59). This was associated with reduced turfgrass quality and verdure. In this study water use was examined through a time course of 1 hour to 96 hours, following mower treatment under controlled environment conditions. Over the course of the study turfs mowed with sharp blades used more water than those with dull. There was a trend of increased water loss initially with dull mower-blade treatments, but this was overcome within 24 hours by sharp mower-blade treatments. *Leaf-tissue mutilation occurring after mowing with a dull mower could cause an increase in water loss per leaf when compared with leaves cut with sharp blades, but this response is most likely overcome on a turfgrass stand basis, since there is a reduction in total leaf area.*

Nutrition. The nutrient-supplying capacity of soils may not be sufficient to meet nutritional needs of turfgrass plants. Turfgrasses need a balanced supply to maintain growth, recuperative potential, and quality. When the soil cannot supply an adequate balance, turfgrass managers must fertilize to maintain desired characteristics. Fertilization influences turfgrass growth and water use. This is particularly true when nitrogen is examined.

Several researchers have reported increased water use rates with increased nitrogen nutrition (7, 11, 19, 27, 36, 56, 64). Water-use rate of 'Penncross' creeping bentgrass grown in controlled environment and treated with nitrogen rates ranging from 0 to 30 g N/m^2 monthly (0 to 6 lb $N/1000$ ft^2 monthly) with phosphorus, potassium, and micronutrients supplied at constant levels was assessed in a Michigan study (56). Water use increased with nitrogen application rate until 10 g N/m^2 (2.0 lb $N/1000$ ft^2) was applied and then declined at rates in excess of this level. Shoot growth, shoot density, and root organic matter production followed trends identical to those observed for water use. Leaf width, shoot density, and shoot growth were positively correlated to water use ($r = 0.85$, $r = 0.91$ and $r = 0.80$, respectively), but root organic matter production was not. Their results demonstrated the importance of nitrogen nutrition on increased plant surface area and increased evapotranspiration rates. The nonsignificant relationship of root organic matter production to water-use rate is not surprising, since there are confounding relationships of root and shoot growth in regard to nitrogen nutrition. In a review of turfgrass literature, it was pointed out that numerous researchers have demonstrated increased root and shoot growth as nitrogen nutrition is raised from a deficiency level, but this trend continues only until a critical growth stimulation is obtained (3). At this critical point and at higher nitrogen levels, turfgrass root growth may stop or be significantly suppressed. Turfs grown under these conditions and exposed to high atmospheric evaporative demand are more prone to wilt than those growing under adequate nitrogen nutrition. Common Kentucky bluegrass turfs fertilized with 10 g N/m^2 (2.0 lb $N/1000$ ft^2) had a greater tendency to wilt and were more prone to drought injury than unfertilized turfs (25). The consumptive water use of warm-season and cool-season turfgrass species grown under two management regimes was studied in Arizona. Low-management turfs received 5 g N/m^2 bimonthly and high-management turfs received 5 g N/m^2 monthly. Turf appearance was characterized under low management as being "of relatively low quality similar to that on many campus, park, and home lawns"; turf appearance under high management was seen as "better than that of many home lawns, but not as lush as many commercial lawns" (26). Consumptive water-use rates for low-management turfs were approximately 340 mm (13.5 inches) per year less than for high-management turfs. Results of the research indicated a considerable potential for water-use reduction by manipulating nitrogen nutrition within the usual maintenance range for desired turfgrass quality (26).

Increasing N-levels from 5.5 g N/m^2 (1.1 lb $N/1000$ ft^2) to 22.5 g N/m^2 (4.5 lb $N/1000$ ft^2) was found to decrease the amount of water required to produce a unit of dry matter in several warm-season forage species (5). This response was measured even in a dry year when grasses treated with high nitrogen rates were observed to wilt more rapidly and apparently use water faster than those receiving low nitrogen treatments. Research using Kikuyugrass (*Pennisetum clandestinum* Hochst.) in Israel demonstrated that dry matter production was not markedly raised by regular nitrogen fertilization unless irrigation was applied frequently (36). Kikuyugrass fertilized with 2.1 kg $N/1000$ m^2 (0.5 lb $N/1000$ ft^2) month and irrigated every 25 days produced no more dry matter than did nonfertilized grass irrigated every seven days. It was concluded that regular nitrogen fertilization would not encourage lush growth provided irrigation was not applied too frequently. This is likely, since nitrogen would be inefficiently used in such a program.

Another study observed the effects of three levels of nitrogen and phosphorus applied in all combinations to a mixed stand of orchardgrass (*Dactylis glomerata* L.), creeping red fescue (*Festuca rubra* L.), and smooth bromegrass (*Bromus inermis* Leyss) on efficiency of water use (27). This work demonstrated that increasing nitrogen rates increased water use in the mixture. Phosphorus applications showed a slight increase, but the trend was not significant and there was no significant effect of the nitrogen × phosphorus interaction. It was suggested that year-to-year variation in weather conditions affected water consumption through evapotranspiration at least as much as the increased plant surface area resulting from nitrogen fertilization. However, water-use efficiency was found to increase with increasing nitrogen input. This occurred because the increased yield from nitrogen fertilization offset the rise in evapotranspiration.

Water-use efficiency, as it relates to nitrogen fertilization in turfs, is an important concept. Turfgrass water-use efficiency relative to nitrogen nutrition is not best discussed in relation to dry matter production or clipping yield. Turfgrass managers are interested in nutritional programs that meet the need of the turfgrass plant but avoid excessive growth and undue stress. Discussion of water-use efficiency as it relates to nitrogen nutrition should center around nitrogen rates that maintain desired turfgrass quality, recuperative rate, and reduced water consumption. In this regard, turfgrass water-use efficiency would be a relative term, relating to desired turfgrass quality and function, and would differ in comparison between a golf green and a utility turf. Judicious and appropriately timed nitrogen applications oriented toward meeting turfgrass nutritional needs should be the goal of managers interested in the most efficient use of available water. Research relating nitrogen carrier, rate, and application timing with drought resistance and tolerance is needed.

In research on the influence of phosphorus and potassium nutrition, in conjunction with nitrogen rate and application timing on Kentucky bluegrass drought recovery, it was found that recovery was greater for turfs treated with fall-applied nitrogen than for spring (53). Phosphorus increased recovery but was interactive with nitrogen rate and application timing; potassium benefited recovery regardless of nitrogen and phosphorus treatments. One study (48) showed that bentgrass produced more root weight with fall nitrogen fertilization than spring; these results were confirmed (63) and demonstrated a benefit in root development from foliar-applied iron made in conjunction with late-season nitrogen applications. These iron treatments also reduced desiccation injury associated with nitrogen treatment.

Water use as it relates to potassium nutrition has not been clearly delineated. No data are available in turfgrass literature regarding the direct influence of potassium on turfgrass water use. Researchers have reported increased rooting, both in extent and depth, with increased potassium nutrition (37, 41). Increased rooting associated with potassium treatment could contribute to observed drought resistance and improved drought injury recovery in turfs, particularly those that are deficient in potassium (53, 54). Increased turgor and reduced tissue moisture content were reported with increased potassium fertilization (57). In addition, it was found that the wilting tendency of a 'Fylking' Kentucky bluegrass turf decreased as potassium nutrition was increased from 0 to 40 g K/m² season, but this trend was negated as increasing nitrogen nutrition levels (10 to 40 g N/m²) were superimposed across these rates (55). More research is needed regarding potassium nutrition and its relationship to turfgrass water use, wilting tendency, and drought resistance.

Nutrition has a complicated and confounded role in turfgrass water use. Its relationship with other cultural practices and its intrarelationship with nutrient interactions make it a difficult and complex research area. Managers should adjust their nutrition programs to produce the least amount of excess topgrowth and the greatest amount of rooting possible, if they wish to minimize unnecessary water use.

Irrigation. In many areas where turfgrasses are cultured, natural precipitation must be supplemented with irrigation to provide desired turfgrass quality. Irrigation, as a cultural practice itself and interacting with other cultural practices, can affect turfgrass water use (10). Irrigation or the lack of it can influence turfgrass drought resistance and tolerance.

A commonly accepted turfgrass management recommendation is to practice deep but infrequent watering (21). This recommendation is ambiguous since it does not provide a basis for actual amount or frequency of application. One study concentrated on the preconditioning effects of irrigation frequency on percent vegetative cover and water-use rates of 'Penncross' creeping bentgrass turfs exposed to desiccating conditions in a controlled environment (56). Turfs receiving water when visual wilt was evident used 33 percent less water than those watered three and seven times weekly. The decline in water use was positively correlated ($r = 0.98$) to a decrease in vegetative cover and a subsequent decline in turfgrass quality.

This study showed that adjustment in irrigation frequency can affect water-use requirements of turfs. However, this research demonstrates the need for more precise assessment of irrigation frequency to maintain turfgrass quality as well as reduce water use. It was not possible to avoid losses in vegetative cover, using visual wilt as the assessment for timing irrigation frequency. In an examination of irrigation frequencies in Nevada water was applied every 3, 7, or 10 days to turfs growing on loam or sandy loam soils (67). Water was applied at 5.0 cm (2.0 inches) weekly to meet calculated demand. It was found that Kentucky bluegrass-fine fescue turfs could be maintained with high quality with twice-weekly irrigation, regardless of soil type. Seven-day intervals were adequate for the loam soil, and 10-day intervals produced sparse, low-quality turfs.

In one study (38, 74) tensiometers and pan evaporation were used to guide scheduling of irrigation for warm-season and cool-season turfgrasses. Tensiometer and evaporation pan scheduling was compared to a control that depicted typical practices of local turfgrass managers with manual irrigation systems. Turfs receiving irrigation scheduled by tensiometers or evaporation pan received less water than did the control and these treatments followed changing weather patterns more closely than did the control, which used visual water-need estimations. For the warm-season grasses (St. Augustinegrass and bermudagrass) there were no differences in turfgrass quality due to treatment, even though mean-water application rates for tensiometer-scheduled irrigation used as much as 55 percent less water than the control treatment. However, in the second study using cool-season species (tall fescue and Kentucky bluegrass) turfgrass quality was affected by irrigation scheduling. Kentucky bluegrass turfs, scheduled to receive irrigation when tensiometer readings reached 55 cb, maintained poor quality throughout the study, having low shoot density and increased disease incidence. Turf quality of the tall fescue was reduced by this treatment in the first year of testing, but in subsequent years, after the root system had apparently fully developed, there was no reduction in turfgrass quality from this treatment, and all treatments produced comparable results.

Research in Florida using microswitching tensiometers and electronic soil-moisture-sensing devices for control of irrigation scheduling showed reduced water application by 89 percent over conventional approaches without reducing turfgrass quality (2). A study of Kentucky bluegrass responses to limited water showed that with slight nitrogen deficiency bluegrass turf quality declined linearly with decreased water over a range of 100 to 40 percent maximum evapotranspiration (10). With adequate nitrogen nutrition quality was maintained until 70 percent of maximum evapotranspiration was obtained. This appeared to indicate that there was an apparent nitrogen × mowing × irrigation interaction on Kentucky bluegrass turfgrass quality.

A study of consumptive water use of subirrigated turfgrasses indicated that they followed Class A pan evaporation closely during periods of high water demand and active growth (26). Consumptive water use, expressed as a percentage of Class A pan evaporation, ranged from 42 to 80 percent, depending upon the turfgrass species and intensity of management. It was concluded that watering levels in excess of 50 to 80 percent of Class A pan evaporative losses for warm-season and 60 to 85 percent for cool-season turfgrasses would be a wasteful and unnecessary use of water. These results concur closely with other investigations of sprinkler irrigation procedures (6, 26).

In a comparison of tensiometer-controlled irrigation on Kentucky bluegrass to a set schedule, the tensiometer-controlled irrigation reduced water application by 28 to 48 percent over a 4-month period without reducing turfgrass quality (43). The most dramatic savings (66 percent) occurred in fall, when evapotranspirational demand was reduced, and the least savings (11 percent) occurred in August, when evapotranspiration rates were extremely high. It was concluded that the set irrigation schedule was inefficient and higher evapotranspiration losses occurred with turfs receiving more water.

Adjusting irrigation frequency, using electronic moisture-sensing devices, tensiometers, Class A pan evaporation, or combinations of these, will effectively reduce unnecessary applications of water. The previously discussed research verifies this approach across a wide range of growing conditions and turfgrass species. Turfgrass managers interested in incorporating these devices in their irrigation practices should approach them with caution, utilizing local research data and information and on-site experimentation, before embarking on irrigation programming that relies on such devices.

Other irrigation practices may indirectly influence turfgrass water-use rate and water-use efficiency. The practice of syringing may be one of these. However, no data are available regarding syringing and water use. Temperature reduction in a 'Toronto' creeping bentgrass turf (i.e., mat at 5.0 cm soil depth) was reportedly small, but the syringing treatment prevented the turf from reaching maximum temperature and prevented wilt (14). A North Carolina study in 1982 (12) reported that, in the absence of wilt, 'Penncross' creeping bentgrass canopy temperatures were not significantly reduced by syringing. No reference was made to effects on maximum canopy temperature. The likely effect of such an irrigation practice would be to enhance turfgrass water use, since wilt was prevented, but the subsequent retention of turfgrass quality would offset water use through increased water-use efficiency.

Depth of soil wetting indirectly influences turfgrass water use and directly influences turfgrass drought avoidance. Light water applications that wet only a limited portion of the soil profile encourage shallow rooting. Heavier water applications that thoroughly wet the soil profile enhance deep rooting. As was pointed out in one study (35), turfgrasses with extensive root systems draw moisture from deeper in the soil profile than do turfgrasses with limited root systems. This enhances the turf's ability to avoid moisture stress and wilt, because a deep, extensive root system can draw upon a larger soil volume for water and is better suited to meet evapotranspirational demands than a restricted or limited root system.

Soil cultivation. Turfgrasses are often exposed to traffic and, subsequently, to soil compaction injury. Soil compaction affects soil bulk density, aeration, and water retention; these in turn influence turfgrass growth and development. Researchers have observed reduced root growth and shoot growth as a result of soil compaction (20, 43, 51, 61, 66). One researcher reported slower root growth, increased root maturation and decreased root permeability with soil compaction. Work at Kansas State University with Kentucky bluegrass and perennial ryegrass has demonstrated that compaction altered turfgrass rooting both in extent and distribution (43, 44). This research reported that root viability apparently decreased as a result of low oxygen diffusion rates (ODR) and that critical ODR levels were lower for shoot growth response than for root growth. Thus, shoot growth declined immediately after compaction, but root growth declined and root distribution was altered over a longer time. Water use would be anticipated to decline with reduced shoot growth, and in fact, it was reported that turfgrass water use declined with increasing levels of soil compaction in Kentucky bluegrass turfs (43). Similar results were obtained in work with perennial ryegrass, where water use was assessed based on irrigation requirements determined from tensiometer readings. Another study (61) found soil compaction to increase bulk density, water retention, and soil strength and to decrease aeration porosity. It reported a decline in turfgrass quality, clipping yield, nitrogen-use efficiency, evapotranspiration (i.e., 28 percent reduction), and root growth with increasing compaction treatment. The most detrimental effect of compaction on root growth occurred using a water-soluble N-source at 10 g N/m^2.

Results of soil compaction research at Kansas State University and elsewhere demonstrated that soil compaction affects turfgrass shoot and root growth, turfgrass quality, soil-moisture retention, and turfgrass water use. No data are available relating soil cultivation practices to turfgrass water use, but it stands to reason that soil coring or aerification would increase water use on compacted soil sites, since they would enhance turfgrass rooting, quality, and water-use efficiency.

Chemicals

Various chemicals are used in turfgrass cultural programs. It is reasonable to consider these chemicals and their potential effects on turfgrass water-use rates since they may directly or indirectly influence turfgrass growth, leaf area, rooting, canopy resistance, and water use.

Antitranspirants control transpiration at the leaf-air interface by inducing stomatal closure, covering the mesophyll surface with a thin film (i.e., monomolecular film), or by covering the leaf surface with a water-impervious film. Antitranspirants have serious turfgrass management implications since they have a potential detrimental effect on turfgrass photosynthesis and evapotranspirational cooling. Phenylmercuric acetate (PMA) has been used effectively to induce stomatal closure in nongrass species (60, 62, 68, 75). It has promising potential for reducing water loss from transpiration in turfs, but studies with turfgrass species are limited. One researcher (65) conducted research on 'Penncross' creeping bentgrass and 'Tifway' bermudagrass (*Cynodon dactylon* (L.) Pers. × *C. transvaalensis* Burt-Davy) using five potential antitranspirants: abscisic acid (ABA), B-napthozyacetic acid, Stoma-seal (a mixture of PMA and Aqua-Gro), Aqua-Gro (a soil wetting agent), and Experimental No. 913 (a monoglycerol ester of decinyl succinic acid in Aqua-Gro). An assessment was made of the effectiveness of the antitranspirants by measuring differences in transpiration rate of treated and untreated plants and by whether plants were injured by antitranspirant treatments. Transpiration rate was calculated based on the difference between evapotranspiration from live turf and evaporation from dead turf. It was found that ABA and Experimental No. 913 reduced transpiration on 'Penncross' by 59 percent and 29 percent, respectively, without visual injury or increased leaf temperature. In 'Tifway', ABA at 1×10^{-4}M reduced transpiration by 12 percent and shoot growth rate by 23 percent. The conclusion: ABA was an effective antitranspirant, but its use large scale was questionable due to economics. Further research is suggested using ABA, ABA's analogs, and Experimental No. 913 to discern feasible antitranspirants for use in turf.

Monomolecular films have been used on water reservoirs to minimize water loss from evaporation, and have been used with erratic results to suppress transpiration in plants (1, 42). These materials tend to reduce photosynthesis to a greater extent than transpiration. Thick film coatings that cover leaf surfaces, including stomatal pores, have been used as antitranspirants and winter antidesiccants (8, 39, 73). Most use of these materials has been restricted to ornamentals and detailed information is lacking for turf. Shearman and Beard (unpublished) examined Wiltpruf and Stoma seal applied at manufacturer-recommended rates to 'Penncross' creeping bentgrass turfs. Water loss from treated turfs was not significantly different from that of the untreated control. It was apparent that the materials did not effectively coat turfgrass leaf blades. Film-coating materials are of limited benefit under turfgrass-growing conditions, since it would be difficult to obtain uniform leaf-blade coating and mowing would remove coated leaves and reduce effectiveness of treatments. Some skepticism on the potential benefits of antitranspirants in turf was expressed in research conducted on St. Augustinegrass (23). Results indicated that water use in a well-watered turf was influenced to a greater extent by environmental factors external to the plant than by stomatal apertures. It was felt that manipulation of turfgrass morphology or canopy offered a greater potential for reducing evapotranspiration than did antitranspirants.

Wetting agents have received only limited testing in turf and no data covering their effect on water use exist in turfgrass literature. One study observing the effect of Aqua-Gro on transpiration in antitranspirant studies was conducted on creeping bentgrass and bermudagrass (65). Another (29) reported improved water infiltration on hydrophobic soils treated with nonionic wetting agents. Similar findings have been reported for hydrophobic sands, soils and thatch (46, 47, 70). Reduced evaporation and improved soil-water retention have also been found as a result of wetting agent treatments (15, 28, 33). Wetting agents most likely play a limited and indirect role in turfgrass water use through enhancement of soil wetting, water retention, evaporation, and increased available water for turfgrass use.

Pesticides may affect turfgrass water use and water-use efficiency, both directly and indirectly. For example, PMA has been discussed as a potential antitranspirant, but also has eradicative fungicidal properties. Application of PMA as a fungicide could directly reduce transpiration by influencing leaf-blade stomatal apertures. Conversely, an application of a fungicide could indirectly affect transpiration by increasing or decreasing the turfgrass canopy, the root system, or both. The same case could be made for herbicides and insecticides. A decline in water-use rate and water-use efficiency was reported with increasing siduron application rates in a 'Kentucky 31' tall fescue (*Festuca arundinacea* Schreb.) turf (58). Shoot and root dry matter production decreased with increasing siduron rates and degrees of wilting and drought injury were increased at application rates of 13.6 kg/ha^1 (12.0 lb/acre^1) or greater. In a study conducted on the effects of successive preemergent herbicide applications on Kentucky bluegrass turfs, a reduction in water use associated with prosulfalin and benefin treatments was noted (50). The decline in water use was closely associated with reduced turfgrass density and quality. This study and another (58) demonstrate a potentially detrimental effect on turfgrass water use and water-use efficiency most likely by indirect treatment effects on the turfgrass canopy and root systems. Detrimental effects of pesticide applications on water use apparently decrease water-use efficiency and potential drought avoidance. Since in most cases judicious pesticide applications enhance turfgrass quality and performance, their potential influence on turfgrass water-use efficiency would probably be beneficial.

Based on the influence of mowing height, leaf and shoot extension rate, and canopy resistance on transpiration rates in turf, plant growth regulators (PGR) should be carefully considered for reducing turfgrass water-use rates. Research in this area is extremely limited. At Texas A & M University, field experiments were conducted on 'Texas Common' St. Augustinegrass and 'Tifway' bermudagrass, testing an experimental PGR, flurprimidol (EL-500), in various treatment combinations with and without mowing (24). Transpiration rate was assessed at 15-day intervals over a 100-day period. Leaf-area measurements were made on the 60th day and leaf-area indices were calculated from these data. Significant reductions in evapotranspiration (11 percent to 29 percent reduction) were measured through 14 weeks on both mowed and unmowed plots treated with EL-500. Turfs with the lowest leaf-area indices had the lowest evapotranspiration rates. Shearman assessed evapotranspiration rates in Kentucky bluegrass turfs treated with PP-333, EL-500, and Embark compared with an untreated control (53). Turfs were mowed at 5.0 cm whenever they reached 7.5 cm in height. Water use was assessed with field lysimeters by weighing water loss over a 24-hour period at 14, 28, and 42 days after treatment. PGR treatments reduced water use by 23 to 44 percent through 28 days, when compared to the check. There was no significant difference among treatments after 42 days. EL-500 and PP-333 were more effective in reducing evapotranspiration than Embark. Research (24, 53) has demonstrated that plant growth regulators can be used to reduce evapotranspiration rates effectively in warm- and cool-season turfs. This offers a promising area of research investigating various PGR materials, application rates, application timing and turfgrass species, and cultivar responses.

Summary

Most water loss from turfs occurs as evapotranspiration from turfgrass leaves with only a slight amount resulting from leaf sheaths and stems. As much as 90 percent of the evapotranspiration rate occurs as stomatal transpiration with the remaining 10 percent occurring as evaporation from the plant cuticle and moisture in the thatch and soil. Transpiration rates increase as plant-surface area exposed to desiccating conditions increases and canopy resistance decreases. Turfgrass cultural practices have been demonstrated to influence shoot and root growth, water-use rate, and water-use efficiency. Turfgrass scientists do not fully understand all the implications of cultural practices on turfgrass water use, but information is available for suggested guidelines to reduce water consumption.

As mowing height was increased, turfgrass water-use rate increased. This response occurred as a result of increased leaf area, allowing evapotranspiration, coupled with a deeper, more extensive root system, to increase water absorption and enhance the capabilities to support higher evapotranspiration rates. More frequent mowing reduced turfgrass water use, since leaf extension was kept at a minimum and turfgrass canopy resistance increased. Mowing with a sharp, properly adjusted mower resulted in increased water-use efficiency. Plant growth regulators reduced turfgrass water use in a similar response to that observed with mowing frequency. Water-use efficiency increased with nitrogen nutrition level until a critical point was reached. Critical levels minimally affect shoot density but detrimentally influence the extent and depth of rooting. Wilting tendency and drought injury have been offset by increased potassium levels. Maintenance of adequate turfgrass nutrition with an appropriate balance of nitrogen, phosphorus, potassium, and other essential nutrients was imperative for enhanced water-use efficiency. Frequently irrigated turfs used more water than those less frequently watered. Turfs watered when visual wilt symptoms were evident, compared with automatic watering three times weekly, used 33 percent less water. Irrigation frequency was adjusted with tensiometer readings and with Class A pan evaporation without losses in turfgrass quality; furthermore, water conservation as great as 50 percent occurred, compared with conventional watering programs. Seasonal responses were noted for tensiometer-controlled irrigations, compared with scheduled irrigations. Soil compaction impaired water infiltration and percolation, reducing turfgrass water-use efficiency.

Certain antitranspirants and wetting agents reduced turfgrass evapotranspiration by as much as 59 percent without visual damage or increase in leaf-blade temperature. More research is needed to delineate this response. Other chemicals, such as preemergent herbicides, were reported to influence water-use rates of cool-season turfgrasses, particularly after several yearly applications.

Specific amounts of water conservation resulting from manipulation of various cultural practices on specific turfgrass species and cultivars cannot precisely be stated, since detailed research is not yet available. However, the relative magnitude of response should be comparable. Turfgrass managers should keep in mind that no one cultural practice will reduce water consumption, since neglect of other critical practices can result in a decline of turfgrass quality and reduced water-use efficiency.

Literature cited

1. AUBERTINE, G. M., and G. W. GORSLINE
   1964. Effect of fatty alcohol on evaporation and transpiration. Agron. J. 56:50–52.
2. AUGUSTINE, B. J., G. H. SNYDER, and E. O. BURT
   1981. Turfgrass irrigation water conservation using soil moisture sensing devices. Agron. Abstracts. p. 123.
3. BEARD, J. B.
   1973. Turfgrass: Science and Culture. Prentice-Hall, Inc. Englewood Cliffs, N.J. 658 pp.
4. BIRAN, I., B. BRAVDO, I. BUSHKIN-HARAV, and E. RAWITZ
   1981. Water consumption and growth rate of 11 turfgrasses as affected by mowing height, irrigation frequency, and soil moisture. Agron. J. 73:85–90.
5. BURTON, G. W., G. M. PRINE, and J. E. JACKSON
   1957. Studies of drought tolerance and water use of several southern grasses. Agron. J. 49:498–503.
6. BUSCH, C. D., and W. R. KNEEBONE
   1966. Subsurface irrigation with perforated plastic pipe. Trans. ASAE 9:100–101.
7. CARROLL, J. C.
   1943. Effects of drought, temperature, and nitrogen on turfgrasses. Plant Phys. 18:19–36.
8. COMAR, C. L., and C. G. BARR
   1944. Evaluation of foliage injury and water loss in connection with use of wax and oil emulsions. Plant Phys. 19:90–104.
9. CRIDER, F. J.
   1955. Root growth stoppage resulting from defoliation of grass. U.S. Tech. Bull. No. 1102. 23 pp.
10. DANIELSON, R. E., C. M. FELDHAKE, and J. D. BUTLER
    1981. Limited evapotranspiration by turfgrass management under water deficiencies. Agron. Abstracts. p. 124.
11. DEXTER, S. T.
    1937. The drought resistance of quackgrass under various degrees of fertilization with nitrogen. J. Am. Soc. of Agron. 29:568–576.
12. DIPAOLA, J. M.
    1982. The influence of syringing on the canopy temperatures of bentgrass greens. Agron. Abstracts. p. 141.
13. DOSS, B. D., O. L. BENNETTE, D. A. ASHLEY, and H. A. WEAVER
    1962. Soil moisture regime effect on yield and evapotranspiration from warm season perennial forage species. Agron. J. 54:239–242.
14. DUFF, D. T., and J. B. BEARD
    1966. Effects of air movement and syringing on the microclimate of bentgrass turf. Agron. J. 58:495–497.
15. ENGEL, R. E., and R. B. ALDERFER
    1967. The effect of cultivation, topdressing, lime, nitrogen and wetting agent on thatch development in 0.25 inch bentgrass turf over a ten-year period. 1967 Report on Turfgrass Research at Rutgers University. N.J. Agric. Exp. Sta. Bull. 818, pp. 32–45.
16. ERIE, L. J., O. F. FRENCH, and K. HARRIS
    1965. Consumptive use of water by crops in Arizona. Univ. of Ariz. Agric. Exp. Sta. Bull. 169.
17. EVANS, M. W.
    1949. Kentucky bluegrass. Ohio Agric. Exp. Sta. Bull. No. 681. 39 pp.
18. EVERSON, A. C.
    1966. Effects of frequent clipping at different stubble heights on western wheatgrass (*Agropyron smithii* Rybd.) Agron. J. 58:33–35.

19. FELDHAKE, C. M., R. E. DANIELSON, and J. D. BUTLER
    1981. Maximum evapotranspiration by turfgrass-environmental and management factors. Agron. Abstracts. p. 125.
20. GORE, A. J. P., R. COX, and T. M. DAVIES
    1979. Wear tolerance of turfgrass mixtures. J. Sports Turf Res. Inst. 55:45–68.
21. HAGAN, R.
    1955. Watering lawns and turf and otherwise caring for them. pp. 462–467. In A. Stefferud (ed.), Water, The Yearbook of Agric. USDA.
22. HART, R. H., and G. W. BURTON
    1966. Prostrate vs. common dallisgrass under different clipping frequencies and fertility levels. Agron. J. 58:521–522.
23. JOHNS, D.
    1980. Resistance to evapotranspiration from St. Augustinegrass [Stenotaphrum secundatum (Walt.) Kuntze] turf. Ph.D. Thesis. Texas A & M Univ., College Station, Tex. 88 pp.
24. JOHNS, D., and J. B. BEARD
    1982. Water conservation—a potentially new dimension in the use of growth regulators. Texas Turfgrass Research—1982. Texas A & M Univ., College Station, Tex. pp. 41–42.
25. JUSKA, F. V., and A. A. HANSON
    1967. Effect of nitrogen sources, rates and time of application on the performance of Kentucky bluegrass turf. Proc. Am. Soc. Hort. Sci. 90:413–419.
26. KNEEBONE, W. R., and I. L. PEPPER
    1982. Consumptive water use by subirrigated turfgrasses under desert conditions. Agron. J. 74:419–423.
27. KROGMAN, K. K.
    1967. Evapotranspiration by irrigated grass as related to fertilizer. Canadian J. of Plant Sci. 47:281–287.
28. LAW, J. P.
    1964. The effect of fatty alcohol and nonionic surfactant on soil moisture evaporation in controlled environment. Soil Sci. Soc. Am. Proc. 28:695–699.
29. LETEY, J., N. WELCH, R. E. PELISHEK, and J. OSBORN
    1963. Effect of wetting agents on irrigation of water repellent soils. Turfgrass Culture. 13(1):1–2.
30. MADISON, J. H.
    1960. Mowing of turfgrass I. The effect of season, interval and height of mowing on the growth of Seaside bentgrass turf. Agron. J. 52:449–452.
31. ———
    1962a. Mowing of turfgrass II. Responses of three species of grass. Agron. J. 54:250–253.
32. ———
    1962b. Turfgrass ecology. Effects of mowing, irrigation, and nitrogen treatments of *Agrostis palustris* Huds.; 'Seaside' and *Agrostis tenuis* Sibth., 'Highland' on population, yield, rooting and cover. Agron. J. 54:407–412.
33. ———
    1966. Effects of wetting agents on water movement in the soil. Agron. Abstracts. p. 35.
34. ———
    1971. Principles of Turfgrass Culture. Van Nostrand Reinhold Co. New York, N.Y. 420 pp.
35. MADISON, J. H., and R. M. HAGAN
    1962. Extraction of soil moisture by Merion bluegrass (*Poa pratensis* L. 'Merion') turf, as affected by irrigation frequency, mowing height and other cultural operations. Agron. J. 54:157–160.
36. MANTELL, A.
    1966. Effect of irrigation frequency and nitrogen fertilization on growth and water use of Kikuyugrass lawn (*Pennisetum clandestinum* Hochst.). Agron. J. 58:559–561.
37. MARKLUND, F. E., and E. C. ROBERTS
    1967. Influence of varying nitrogen and potassium levels on growth and mineral composition of *Agrostis palustris* Huds. Agron. Abstracts. p. 53.
38. MARSH, A. W., R. A. STROHMAN, S. SPAULDING, V. YOUNGNER, and V. GIBEAULT
    1980. Turfgrass irrigation research at the University of California. Irrig. J. July/August 1980:20–21, 32–33.
39. MARSHALL, H., and T. E. MAKI
    1946. Transpiration of pine seedlings as influenced by foliage coatings. Plant Phys. 21:95–101.
40. MITCHELL, K. J., and J. R. KERR
    1966. Differences in rate and use of soil moisture by stands of perennial ryegrass and white clover. Agron. J. 58:5–8.
41. MONROE, C. A., G. D. COORTS, and C. R. SKOGLEY
    1969. Effects of nitrogen-potassium levels on the growth and chemical composition of Kentucky bluegrass. Agron. J. 61:294–296.
42. OERTHI, J. J.
    1963. Effects of fatty alcohols and acids on transpiration of plants. Agron. J. 55(2):137–138.
43. O'NEIL, K. J., and R. N. CARROW
    1982. Kentucky bluegrass growth and water use under different soil compaction and irrigation regimes. Agron. J. 74:933–936.
44. O'NEIL, K. J., and R. N. CARROW
    1983. Perennial ryegrass growth, water use and soil aeration status under soil compaction. Agron. J. 75:177-180
45. OSWALT, D. T., A. R. BERTRAND, and M. R. TEEL
    1959. Influence of nitrogen fertilization and clipping on grass roots. Soil Sci. Soc. Am. Proc. 23:228–230.
46. PAUL, J. L., and J. M. HENRY
    1973. Nonwettable spots on greens. Proc. Calif. Golf Course Supt. Inst., Univ. of Calif. Davis. p. 65.
47. PELISHEK, R. E., J. OSBORN, and J. LETEY
    1962. The effect of wetting agents on infiltration. Soil Science Soc. Am. Proc. 26:595–598.
48. POWELL, A. J., R. E. BLASER, and R. E. SCHMIDT
    1967. Effect of nitrogen on winter root growth of bentgrass. Agron. J. 59:529–530.
49. PRINE, G. M., and G. W. BURTON
    1956. The effect of nitrogen rate and clipping frequency upon yield, protein content, and certain morphological characteristics of coastal bermudagrass (*Cynodon dactylon* (L.) Pers.) Agron. J. 48:296–301.
50. REIERSON, K. A.
    1979. The effects of successive applications of prosulfalin on Kentucky bluegrass turfs. M.S. Thesis. Univ. of Nebr., Lincoln, Nebr. 40 pp.
51. RIMMER, D. L.
    1979. Effects of increasing compaction on grass growth in colliery spoil. J. Sports Turf Res. Inst. 55:153–162.
52. ROBERTSON, J. H.
    1933. Effect of frequent clipping on the development of certain grass seedlings. Plant Phys. 8:425–447.
53. SCHMIDT, R. E., and J. M. BREUNINGER
    1981. The effects of fertilization on recovery of Kentucky bluegrass turf from summer drought. pp. 333–341. In R. W. Sheard (ed.), Proc. of the Fourth Int. Turf. Research Conf. Univ. of Guelph. Guelph, Ontario, Canada.
54. SHEARMAN, R. C.
    1982. Nitrogen and potassium nutrition influence on Kentucky bluegrass turfs. Proc. Seventh Nebraska Turfgrass Field Day and Equipment Show. Univ. of Nebr., Lincoln, Nebr. Dept. of Horticulture Publication No. 82–2:67–70.
55. ———
    1982. Turfgrass responses to plant growth regulators. Univ. of Nebr. Turfgrass Research Report. Department of Horticulture. Lincoln, Nebr. No. 82–1:34–44.
56. SHEARMAN, R. C., and J. B. BEARD
    1973. Environmental and cultural preconditioning effects on the water-use rate of *Agrostis palustris* Huds., cultivar Penncross. Crop Sci. 13:424–427.
57. ———
    1975. Influence of nitrogen and potassium on turfgrass wear tolerance. Agron. Abstracts. p. 101.
58. SHEARMAN, R. C., E. J. KINBACHER, and K. A. REIERSON
    1980. Siduron effects on tall fescue (*Festuca arundinacea*) emergence, growth and high temperature injury. Weed Sci. 28:194–196.
59. SHEARMAN, R. C., D. H. STEINEGGER, T. P. RIORDAN, and E. J. KINBACHER
    1980. Mowing height, mowing frequency, and mower sharpness influ-

ence on Kentucky bluegrass (*Poa pratensis* L.) turfs. Turfgrass Research Summary–1980. Univ. of Nebr., Department of Horticulture Progress Report No. 81-1:57–64.

60. SHIMISHI, D.
    1963. Effect of chemical closure of stomata on transpiration in varied soil and atmospheric environments. Plant Phys. 38:709–712.
61. SILLS, M. J., and R. N. CARROW
    1983. Turfgrass growth, N-use, and water use under soil compaction and N-fertilization. Agron. J. 75:488–492.
62. SLATYER, R. O., and J. F. BIERHUIZEN
    1964. The effect of several foliar sprays on transpiration and water use efficiency of cotton plants. Agric. Meter. 1:42–53.
63. SNYDER, V., and R. E. SCHMIDT
    1974. Nitrogen and iron fertilization of bentgrass. pp. 176–185. *In* E. C. Roberts (ed.), Proc. 2nd Turfgrass Res. Conf. Blacksburg, Va. June, 1973. Am. Soc. of Agron. Madison, Wis.
64. SPRAGUE, V. G., and L. F. GRABER
    1938. The utilization of water by alfalfa (*Medicago sativa*) and by bluegrass (*Poa pratensis*) in relation to managerial treatments. J. of Am. Soc. of Agron. 30:986–997.
65. STAHNKE, G. K.
    1981. Evaluation of antitranspirants on creeping bentgrass (*Agrostis palustris* Huds., cv. 'Penncross') and bermudagrass (*Cynodon dactylon* (L.) Pers. × *Cynodon transvaalensis* Burtt–Davy, cv. 'Tifway'). M.S. Thesis. Texas A & M Univ., College Station, Tex.
66. THURMAN, P. C., and F. A. POKORNY
    1969. The relationship of several amended soils and compaction rates on vegetative growth, root development, and cold resistance of Tifgreen bermudagrass. J. Am. Soc. Hort. Sci. 94:463–465.
67. TOVEY, R., J. S. SPENCER, and D. C. MUCKEL
    1969. Turfgrass evapotranspiration. Agron. J. 61:863–867.
68. WAGGONER, P. E., J. L. MONTEITH, and G. SZEICZ
    1964. Decreasing transpiration of field plants by chemical closure of stomata. Nature. 201:97–98.
69. WELTON, F. A., and J. D. WILSON
    1938. Comparative rates of water loss from sod, turf, and water surfaces. Ohio Agricultural Experiment Station Bi-monthly Bull. 190:13–16.
70. WILKINSON, J. F., and R. H. MILLER
    1978. Investigation and treatment of localized dry spots on sand golf greens. Agron. J. 70:299–304.
71. WILSON, J. D.
    1927. The measurement and interpretation of the water-supplying power of the soil with special reference to lawn grasses and some other plants. Plant Phys. 2:384–440.
72. WOOD, G. M., and J. A. BURKE
    1961. Effect of cutting height on turf density of Merion, Park, Delta, Newport, and common Kentucky bluegrass. Crop Sci. 1:317–318.
73. WOOLEY, J. T.
    1967. Relative permeabilities of plastic films to water and carbon dioxide. Plant Phys. 42:641–643.
74. YOUNGNER, V. B., A. W. MARSH, R. A. STROHMAN, V. A. GIBEAULT, and S. SPAULDING
    1981. Water use and turf quality of warm-season and cool-season turfgrasses. pp. 251–257. *In* R. W. Sheard (ed.), Proceedings of the Fourth International Turf. Research Conf. Univ. of Guelph. Guelph, Ontario, Canada.
75. ZELITCH, I., and P. E. WAGGONER
    1962. Effect of chemical control of stomata on transpiration and photosynthesis. Proc. Natl. Academy of Sci. 48:1101–1108.

Practicum

7 Influence of water quality on turfgrass

JACK D. BUTLER

Jack D. Butler received his Ph.D. in horticulture from the University of Illinois. For 10 years he was employed by the University of Illinois as Extension turfgrass specialist. In 1971 he moved to Colorado State University where he serves as professor (turfgrass) and Extension turf specialist.

PAUL E. RIEKE

Paul E. Rieke received his B.S. degree from the University of Illinois and his Ph.D. degree from Michigan State University. Currently, he is professor of turfgrass management at Michigan State University, where his emphasis is on sod production and turfgrass fertility.

DAVID D. MINNER

David D. Minner received a B.S. in plant science from the University of Delaware, an M.S. in agronomy from the University of Maryland, and a Ph.D. in horticulture from Colorado State University. Currently, he is an assistant professor of horticulture at the University of Missouri.

*The availability of **adequate** water—rather than **water quality**—receives primary consideration before turfgrass sites are developed. This oversight can cause serious problems. Seawater intrusion into wells and lakes, tidal flow into streams, brackish water, runoff water where soils are saline, effluent water, and wells in arid and semiarid regions are frequent causes of low water quality.*

Water-quality concern includes chemical, physical, and biological factors. From an irrigation standpoint, chemical composition normally receives the most attention. An appropriate and accurate water test can determine the chemical quality of irrigation water. An irrigation water test usually provides information on electrical conductivity (EC), sodium adsorption ratio (SAR), pH, cations, anions, salinity hazard, and boron content.

Total dissolved solids (TDS) or EC normally receive the most attention from an irrigation water-quality standpoint. Based on TDS or EC, irrigation water quality can be broadly classified for suitability for specific use.

High levels of potassium are seldom of concern in irrigation water. In some instances potassium in sewage effluent water may be adequate to meet turf needs. The sodium hazard of irrigation water is of some concern because of possible toxicity to plants, but the major concern is with the effects on soil. The pH of water is routinely determined, but unless it is exceptionally high or low, it does not pose major restrictions on the use of water for irrigation.

Calcium and magnesium cause varying concerns when present in water in high amounts. Saline soils that result from excessive calcium and magnesium deposited from irrigation can decrease plant growth.

Bicarbonates in water can cause serious problems because they can raise the SAR of the soil. Chlorides adversely affect many ornamental plants. On poorly drained soils, excesses of nitrate may result from irrigation, especially irrigation with effluent water. Under most situations the presence of nitrogen in the water would be advantageous, and in some situations its presence may need to be considered in developing turf fertilizer programs.

Occasionally boron levels in irrigation water may be high enough to cause turfgrass damage. Because boron accumulates in leaf tips, removal by mowing may allow tolerance to fairly high levels.

Biological attributes of irrigation water have received limited attention. Organic components of water may reduce the level of available oxygen in the soil.

Suspended solids in irrigation water can cause serious irrigation equipment problems as well as silting in storage facilities.

Three important ways of handling water of poor chemical quality involve drainage and salt leaching, blending water of different quality, and growing grasses that tolerate salt-laden soils. Drainage improvements—including flexible drain tubes and soil amendment with porous materials—may be used to manage salt-contaminated water better. Leaching requirement equations assist in the management of salt-laden soils. Water quality is an important factor in determining water amounts for leaching salts. A simplification of leaching requirements for saline soils indicates that an acre-foot per acre of good-quality water passing through a foot of soil will reduce salinity by about 80 percent.

Soils with high sodium levels require special treatment. Gypsum, sulfur, and sulfuric acid are among the materials used to treat sodic soils. The amounts of materials to apply for treatment of sodic soils are best determined by soil testing.

Salt-tolerant grasses may be used to vegetate salt-affected areas. The benefits of using salt-tolerant grasses may be short term if salts continue to accumulate through irrigation. A more permanent solution to salt problems normally results from drainage improvements. Weeping alkaligrass, Lemmon alkaligrass, 'Seaside' creeping bentgrass, bermudagrass, *Paspalum vaginatum*, and St. Augustinegrass have good salt tolerance. Kentucky bluegrass and colonial bentgrass have poor tolerances. A wide range of salinity tolerance may occur between cultivars of the same species.

7 Influence of water quality on turfgrass

Jack D. Butler, Paul E. Rieke and David D. Minner

In turfgrass management water quality, as well as quantity, is frequently of major concern. The demand that turfgrasses now be produced in greater quantities near oceans, in arid and semi-arid regions, and where water pollution is common causes even more concern about water quality. On many sites lack of good-quality water limits establishment and maintenance of suitable turfgrass.

The information presented here is directed to water quality concerns as they relate to turfgrass production. The reader is referred to a wide variety of references (4, 37, 41, 55, 64, 71) for more detailed treatment of this complex subject.

Water quality, especially as it relates to dissolved salts, influences soil quality. Frequently, soil chemical quality and physical quality are determined by testing before growing crops. Testing soil rather than water has certain advantages. Soil tests can determine the potential for leaching salts as well as provide information on levels of existing salts and trace elements, often deposited by irrigation water, as they relate to specific crop production.

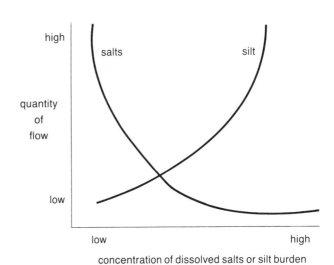

FIGURE 1. A generalized relationship of dissolved mineral matter and suspended material with river-flow quantity (21)

Common sources of water

Meteorological water. Water quality requirements for irrigation are the same, regardless of the source of the water supply (38). The most common water utilized for turfgrass production is meteorological, both rain and snow, in origin. Usually it is assumed that these waters are pure, and from a turfgrass standpoint they usually are. Natural waters contain low levels of dissolved minerals, as well as gaseous matter. Generally, the total salt concentration of rainwater decreases with distance from the sea (37). The ionic ratios for rainwater near the sea are similar to those of the sea, but they differ greatly as distance from the sea increases. Heavy metals may be found in measurable amounts in precipitation (38); several of these (cadmium, lead, mercury, etc.) have been used at heavy rates as turfgrass pesticides with only limited difficulties. Consequently, amounts found in precipitation would be of minor concern, from a turfgrass standpoint. Acid rainfall has become a major environmental concern. Unpolluted natural precipitation is often considered to be slightly acidic with a pH of 5.65. Acid rain has come to mean rainfall with a pH below this (53). With turfgrass production at this time, acid rain seems to be of minor consequence; in some instances it may be beneficial in lowering the pH of alkaline soils.

River and stream water. These are frequently used for turfgrass irrigation. It has been suggested that since water is known as the "universal solvent," it follows that runoff contains anything that water will dissolve as it flows (38). Also, undissolved matter may be found in surface-flowing water. Flow rate relates to the purity of river flow. Figure 1 represents the effect of river-flow quantity on the concentration of dissolved salts and suspended solids (21).

Lakes and ponds. These are frequently used for storage of water for irrigation purposes, may be sources of contamination, or they may allow otherwise high-quality water to become contaminated. Impounded water, although it may supply a fairly constant quantity, can have quality differences that vary with season and depth. Evaporation significantly affects the volume reduction of water; consequently, this tends to concentrate salts dissolved in the water. This phenomenon has the greatest effect on small lakes and reservoirs (21).

Spring water. It is often quite turbid (clouded with sediment) during times of low precipitation and, depending upon the aquifer, it may contain high amounts of dissolved salts.

Ocean water. It frequently contaminates fresh water. Contamination may occur in several ways: by tidal flow, by wind-blown spray, and by intrusion along pervious subsurface layers to contaminate impounded as well as groundwater supplies. Brackish water along seacoasts frequently causes turf-growing problems. Twenty-four ions can reportedly be found in seawater (25). Although the published composition of seawater varies slightly, table 1, page 74, is representative of the major constituents.

It has been pointed out that brackish waters may vary greatly in salt concentration, but the relative proportion of ions remains essentially the same (22). Consequently, the sodic nature of these waters makes them of special concern from a turf production standpoint.

TABLE 1 Seawater composition with SAR = 58 (21)

Cation	ppm	%*	Anion	ppm	%
Na+	10,556	30.61	Cl⁻	18,980	55.05
Mg++	1,272	3.69	$SO_4^=$	2,649	7.68
Ca++	400	1.16	HCO_3	140	0.41
K+	380	1.10	Br	65	0.19

*Total dissolved substances.

Groundwater. It may be constant in quantity, and it is usually constant in quality (38). Groundwater often contains excessive amounts of many different cations and anions. In addition, salts may be contributed in considerable amounts to soils by high water tables. It has been concluded that a practical "critical" water table level is 150 to 200 cm (5.0 to 6.7 ft.) (23).

Water treatment

In addition to these naturally occurring problems, water quality may be chemically altered in treatment for human consumption or aquatic weed control. Chlorination of domestic water for disinfection is common. Chlorine is used at very low levels for treating water. It has been noted that for swimming pool water chlorine residual should be kept between 0.4 and 1.0 ppm (11). For human consumption there is no definite information on amounts to avoid, but concentrations much in excess of 1 ppm have been used with no ill effects. Land plants (turfgrass) are not injured by chlorine concentrations up to 50 ppm (11).

Copper treatment of impounded water, for algae control, is common. Research on bentgrass determined that use of irrigation water with 1.0 ppm Cu from both an inorganic (copper sulfate) and organic (copper-triethanolamine complex) source presented a low probability of turfgrass injury (29). In this work several aquatic herbicides, ranging from a low to high phytotoxicity hazard, were screened. Differences occurred between different formulations of 2,4-D at the same active-ingredient rate, and high hazards occurred at rates tested with simazine and dichlobenil.

Effluent water

Irrigation of turfgrass with effluent water is now common in arid and semiarid regions, and for disposal of an unwanted product in more humid areas. Treated effluent quality, based upon irrigation water criteria, has been found to be only slightly poorer than its municipal source (31). Also, as a general rule, use of water by a city adds 100 to 300 ppm salts; nitrogen content is usually from 20 to 35 ppm. This is often adequate to meet turf requirements when effluent is the sole source of irrigation water.

Of course, wastewater varies greatly in quality. It has been noted that water derived primarily from domestic use will contain little that may be phytotoxic, but if industry has made a major contribution of toxic substances, especially heavy metals, these may present a particular danger (72). Furthermore, the salt load of a typical urban effluent has been characterized as containing the following elements in the following concentrations given as mg/l: Na-70, K-10, Ca-15, Mg-7, Cl-75, SO_4-30, Si-15, PO_4-25, NH_4-20, NO_3 and NO_2-1 (6). The typical salt load of this effluent is 0.6 mmho/cm with an SAR of 8.9. If the original source water has a high salt load (for instance, Colorado River water at 800 ppm TDS), the problem can be significant.

Water quality parameters

As noted, there are various water-quality parameters. Of primary concern are the chemical, physical, and biological factors. From an irrigation standpoint the chemical aspect, i.e., dissolved salts, of water quality has received the most attention.

Chemical quality. To determine the chemical quality of water that will be used to irrigate turfgrass, appropriate, accurate testing is needed. As with soil, representative sampling is essential. A study of sampling techniques pointed out that the mechanics of sampling depend upon use of the information, chemical characteristics, and nature of the supply (71). Samples of irrigation water, collected in plastic or glass bottles, should be analyzed as soon as possible to limit possible chemical changes. As with any sample, positive identification and pertinent information on specific needs should accompany the sample to the testing laboratory.

The constituents determined in analyzing irrigation water are fairly standardized. Such tests determine total concentration of dissolved material as well as those of the more important constituents contributing to the total. The following results are usually reported in testing irrigation water: electrical conductivity, sodium adsorption ratio, pH, cations (sum of cations), anions (sum of anions), salinity hazard, and boron.

In addition, SO_4 sulfur and NO_3 nitrogen may be reported in pounds per acre-foot. Sophisticated equipment used in many water-testing laboratories can easily provide other information on concentrations of micronutrients, such as zinc, copper, and manganese, as well as nickel, lead, cadmium, and arsenic. These are frequently of much more concern for food crops than for turfgrass.

Electrical conductivity (EC). It is measured as resistance with an alternating current bridge. ECe is an expression of the conductivity of the soil solution at saturation (reported in mmhos/cm or mmho/cm $\times 10^{-3}$). The basic EC unit expression for water quality is μmho/cm (micromhos/centimeter). The proper metric unit for expressing EC is decisiemens per meter (dS/m), and one dS/m equals one mmho/cm.

Total dissolved solids (TDS). In addition to the use of EC as an expression of salinity, TDS is used. For most purposes mmho/cm can be converted to TDS (expressed in mg/l or ppm) by multiplying by 640. This is only an approximation since the value 640 can range from 550 to 700, due to the type and quantity of dissolved salt (51).

Various classifications of water have been proposed, based upon EC. USDA Handbook 60 notes conductivity dividing points between classes being 100 to 250 (low), 250 to 750 (medium), 750 to 2,250 (high), and 2,250 to 5,000 (very high) μmho/cm at 25°C. These class limits were selected according to the relationship of the EC of irrigation waters and the EC of saturation extracts of soil. Another commonly used expression of hazards based on the salinity of water is provided in Table 2 (20).

TABLE 2 Salinity hazard of irrigation water (20)

Hazard	Dissolved salt content	
	(ppm)	(EC × 10⁶)
1. Waters for which no detrimental effects will usually be noticed	500	750
2. Waters which may have detrimental effects on sensitive plants	500–1000	750–1500
3. Waters that may have adverse effects on many plants and require careful management practices	1000–2000	1500–3000
4. Waters that can be used for salt tolerant plants on highly permeable soils with careful management practices and only occasionally for more sensitive plants	2000–5000	3000–7500

Sodium adsorption ratio (SAR). The sodium hazard of irrigation water is usually expressed as the proportion of Na$^+$ to Ca^{++} plus Mg^{++} in the water. The formula for calculating SAR is:

$$SAR = \sqrt{\frac{Na^+}{\frac{Ca^{++} + Mg^{++}}{2}}}$$

Ions in this equation are expressed in milliequivalents per liter (Meq./1 × Eq. wt. = ppm) (40).

The sodium hazard of irrigation water is of some concern because of the direct effect (toxicity) to sensitive plants; however, the major concern is with effects on the soil. The diagram in USDA Handbook 60 classifies water with respect to SAR. Specific classifications presented are:

Low-sodium water—Water that can be used to irrigate almost any soil without danger of developing harmful levels of exchangeable sodium (ES).
Medium-sodium water—This water will present an appreciable sodium hazard in fine-textured soils with a high cation exchange capacity (CEC), especially under low-leaching conditions, unless gypsum is present. Water of this quality may be used on soils with good permeability.
High-sodium water—Such water may produce harmful levels of ES in most soils, and it requires special soil management. Gypsum-containing soils may not develop harmful levels of ES from this water. Chemical amendments (such as gypsum) may be required for replacement of ES.
Very high-sodium water—This very poor-quality water is generally unsatisfactory for irrigation, except at low and perhaps medium salinity, where Ca from the soil or applied gypsum or other amendments may make use of such water feasible.

The nomogram (fig. 2) in the handbook may be used to determine the SAR values of irrigation water and to estimate the corresponding ESP values.

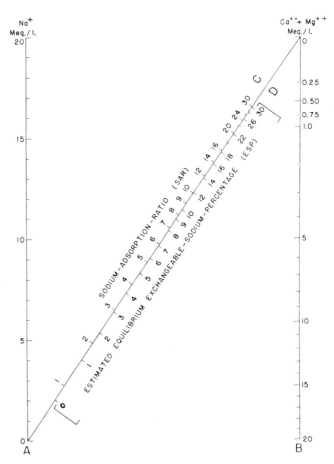

FIGURE 2. Nomogram for determining the SAR value of irrigation water and for estimating the corresponding ESP value of a soil that is at equilibrium with the water (64)

pH. Although pH of water is routinely determined in water testing there is a dearth of information in the literature on its effects on crop production. A strongly alkaline water will have a high pH, relatively high levels of Na$^+$ and HCO$_3^-$, and relatively low concentrations of Ca^{++} and Mg^{++} (71). Water pH may significantly affect plant growth (41). In developing general parameters for field application of pulpmill effluents, researchers noted that the pH should be from 6.5 to 9.0 (7). In unpolluted water pH is seldom low enough to be a problem.

Cations. Calcium, magnesium, sodium, and potassium cause varying concerns when present in irrigation water. Calcium and magnesium effects on the soil can be beneficial, but saline soils, often characterized by a white crust on the surface, can decrease growth of most plants (55). Soil analysis, rather than visual observations, is needed to determine salinity levels properly. Although both Ca and Mg are needed in rather large quantities for plant growth, their detrimental effects when present at excessive levels is of concern. Turfgrasses vary in abilities to tolerate salts in the soil. (See salt-tolerant grasses.)

Sodium. Influences of sodium in water, as noted, are primarily upon soil physical characteristics. This then can complicate the determination of direct plant injury caused by sodium. Most reported instances of sodium injury to plant tissue have been with woody plants, specifically almond and avocado (64). Sodium contents of 0.25 to 0.50 percent on a plant tissue dry weight basis may cause specific injury to sweet gum, avocado, stone fruits, and citrus (51).

Potassium. USDA Handbook 60 notes that the occurrence of high concentrations of potassium in soil solution is rare. The value of K in sewage effluent has been noted (32). Concentrations of primary nutrients (N, P and K) in a typical effluent are low, but continued use at high rates could add significant amounts of fertilizer (6). In most instances, turf and trees obtain all the phosphorus and potassium they need, and a large part of their nitrogen needs from effluent water (24).

Anions. Water high in *bicarbonate* (HCO_3^-) tends to precipitate calcium carbonate ($CaCO_3$) and magnesium carbonate ($MgCO_3$) when soil solutions concentrate through evaporation. Consequently, the SAR value (relative proportion of Na^+ to Ca^{++} and Mg^{++}) increases. This, in turn, increases the sodium hazard of the water to a level greater than indicated by the SAR value (40). This hazard can be evaluated in terms of residual sodium carbonate (RSC). The following formula has been devised (16):

$$RSC = (CO_3^= + HCO_3^-) - (Ca^{++} + Mg^{++})$$

in which concentrations are expressed in meq/l. Water with >2.5 meq/l RSC is probably not suitable for irrigation. Those with 1.25 to 2.5 are marginal, and those containing <1.25 meq/l are probably safe (70).

Chloride. Areas of chloride deficiency are not likely to be found on irrigated soils (66). Furthermore, the nutrient anions, NO_3^-, Cl^-, and $SO_4^=$, are completely mobile in the soil solution because of the absence of strong absorption forces. It has been noted in Handbook 60 that many plant species are no more sensitive to chloride salts than to sulfate salts; there is good evidence, however, for the specific toxicity of chloride to some tree and vine crops.

The accumulation of chloride (0.5 to 1.0 percent on a dry weight basis) may cause leaf burn, premature leaf drop, or stem dieback in many ornamentals; also, salt-sensitive plants may be killed (51). Work with alkaligrass (*Puccinellia* spp.) found that accumulation of cations applied as chloride salts was higher than accumulation of those applied as sulfate salts (27). Also, a 2-month exposure to 84 meq/l or more of any sulfate salt, or 112 meq/l or more of any chloride salt caused death in established stands of both 'weeping' (*P. distans*) and 'Lemmon' (*P. lemmoni*) alkaligrass.

Sulfate. Several means of the origin and accumulation of salt in soil and water have been noted (21). For sulfates these are:

a. 0.08 percent in igneous rock
b. SO_2 and H_2SO_4 from volcanic gasses (about 25 percent)
c. Oxidation of ores, e.g., FeS_2 and native elements
d. Decomposition and oxidation of S in plant material
e. Maritime fogs (50 ppm or more)

Consequently, the potential for high levels of sulfate in cropping systems is great. It is felt that the specific sensitivity of crops to high concentrations of sulfate relates to a tendency for this to limit plant uptake of calcium (64). Working with high levels of sulfate salts has produced sulfate levels as high as 3.25 percent in roots and 2.00 percent in shoots of weeping alkaligrass (27). $SO^=$, as noted, is leachable, and can usually be managed in the root zone.

Nitrate. Supplemental applications of nitrogen-containing fertilizers are commonly practiced on turfgrass installations, because of high plant requirements and the mobility of NO_3^- in permeable soils. Excesses of N may occur on poorly drained sites. Supplemental nitrogen is usually considered a benefit to be derived from irrigation water, especially effluents. One observation of a typical effluent water noted that when 40 acre-inches of water are added, 160 pounds of N per acre would be applied (6). Depending on the kind of grass, quality of turf needed, and location, this might prove sufficient or excessive to meet needs. In a discussion of a large system for effluent use (disposal) it was observed that it was designed for delivery of 600 pounds of N per acre per year (69). However, approximately 95 percent of the N was in organic form, of which only 30 to 40 percent is partially available the year applied. Evidently, for most effluent waters and for certain natural waters, concerns with nitrate content of the water can be important in developing turfgrass fertility programs.

Boron. USDA Handbook 60 notes that boron, an essential element for plant growth, may be found in virtually all natural waters in amounts from traces to several ppm. One classical work (15) with boron is noted.

Because boron is weakly absorbed in soil, marginal levels in irrigation water reportedly may not be immediately toxic; however, continued use exceeding specified levels cannot be tolerated (4). Furthermore, the tolerance levels for boron and other trace elements take into account leaching to some degree since they are based, in part, on field observations of plant response to waters of known toxic element concentration. Boron is difficult to leach, and it takes roughly three times the amount of water required to leach chloride (51). In work with high boron sewage effluent on golf greens, boron concentrations were noted as high as 7.8 ppm in both water and soil (13). A rapid rise of the boron level in the soil was associated with the high boron level of the effluent irrigation water. During the short duration of this work the turf remained in excellent condition.

Little decline in vigor was noted with high boron supplies

when five turfgrasses were clipped frequently. Boron was found to accumulate in tips, and clipping off tips may allow for tolerance of higher boron levels. In general, the order of boron accumulation in six grasses was bermudagrass (*Cynodon dactylon*)<Japanese lawngrass (*Zoysia japonica*)<Kentucky bluegrass (*Poa pratensis L.*)<'Alta' tall fescue (*Festuca arundinaceae*)<perennial ryegrass (*Lolium perenne*)<creeping bentgrass (*Agrostis palustris*) (50).

There have been references to trace elements, above, and in rare instances there may be concern for other heavy metals. The National Academy of Science has recommended that for continuous use irrigation effluent water should contain no more than 0.005 ppm Cd, 0.200 ppm Cu, 0.500 ppm Ni, or 5.000 ppm Zn. Most renovated wastewater from domestic sources will meet these standards (6).

Potable water

Quality standards (Tables 3 and 4) for potable (drinking) water are stringent, when compared to irrigation water quality criteria (63). This is especially true when comparing drinking water standards as they would relate to TDS and to heavy metals used as turf pesticides.

TABLE 3 U.S. Public Health Service chemical standards of drinking water, 1962

Recommended maximum allowable concentrations where other more suitable supplies are, or can be made available:

Substance	Concentration in ppm
Alkyl benzene sulfonate (ABS)	0.5
Arsenic (As)	0.01
Chloride (Cl)	250
Copper (Cu)	1
Carbon chloroform extract (CCE)	0.2
Cyanide (CN)	0.01
Iron (Fe)	0.3
Manganese (Mn)	0.05
Phenols	0.001
Sulfate (SO$_4$)	250
Total dissolved solids (TDS)	500
Zinc (Zn)	5

TABLE 4 U.S. Environmental Protection Agency national interim primary drinking water standards, December 1975

Maximum contaminant level which shall constitute grounds for outright rejection of the supply:

Substance	Concentration in ppm
Arsenic (As)	0.05
Barium (Ba)	1
Cadmium (Cd)	0.010
Total chromium (Cr)	0.05
Fluoride (F)	1.4 to 2.4*
Lead (Pb)	0.05
Mercury (Hg)	0.002
Nitrate (as N)	10
Selenium (Se)	0.01
Silver (Ag)	0.05

*Varies with annual average of maximum daily air temperature.

Biochemical oxygen demand (BOD) and chemical oxygen demand (COD) are normally expressed in mg/l. It has been noted (38) that the BOD of river water will be from 1 to 2 ppm, and dissolved oxygen (D.O.) less than saturation. The primary concern of D.O. is for fish. D.O. has been mentioned as a water quality criterion, and concern is not with the oxygen content of the water *per se*, but rather with the effect that irrigation with a given water will have on available oxygen in the soil for plant growth (41). Furthermore, in using certain effluents for irrigation, the presence of readily reducible organic compounds may result in reducing available oxygen in the soil. BOD or COD values can reportedly be indicative of this, and, also, little is known about BOD value change affected by sprinkler irrigation or infiltration into the soil (41).

Physical quality

The major physical consideration of water quality is with the suspended solid load. Figure 1 (21) relates to stream flow and sediment-carrying ability, but it should be noted that few natural streams are completely free of suspended materials (71). The range of sediment concentrations in a stream during a year are much greater than for dissolved solids. In 90 percent of the country in rivers the prevailing concentrations are 900 ppm dissolved solids and 8000 ppm sediment (52). Municipal waters from western rivers have had turbidities of 30,000 and 50,000 ppm (12). Filters, frequently of sand, sediment basins, and chemical coagulants, may be used to reduce turbidity. Sediment, although creating many problems such as silting in streams and lakes, may be beneficial. The fertile sediment carried by the Nile kept its valley productive for centuries. In some instances water sediment that acts as a channel sealant has been removed. This allowed otherwise impervious soils that contained running water to leak badly from streambeds.

Solving problems associated with poor-quality water

The various ways of solving problems associated with poor-quality water require more indepth discussion. Three important solutions are concerned with drainage and salt leaching, blending of water of different quality, and growing grasses that tolerate salt-laden soils.

Drainage has been an important management practice on many agricultural soils for more than 2,000 years (65). Drainage is the process of removing excess water from an area by surface and subsurface means. Drainage of soils, in addition to its role in salt leaching, results in better aeration, deeper rooting, improved trafficability, potential for reduced compaction, reduced susceptibility to water erosion, lowered disease potential, more rapid warming of soils in spring in colder regions, a broader range of crops that can be grown, and improved crop growth and yields (68). Leaching is considered a negative influence in humid regions since loss of nutrients from the root zone of crops leads to acidification of soils, inefficient fertilizer use, and pollution of groundwater (44).

It is the leaching potential that occurs with drainage that is a positive factor in crop management in arid and semiarid regions where irrigation is common (56). Sources of salt accumulation in the soil include: 1) salts in irrigation water, 2) dissolving of precipitated salts in soils by increased use of water, 3) saline water sources from higher ground, 4) upward movement of salts from a shallow water table, 5) fertilizers and manures used in agricultural production, 6) other waste products applied to the land, 7) weathering of soil minerals, and 8) precipitation (56).

Classification of salt-affected soils. Soils potentially affected by highly soluble salts or sodium have been classified as saline or saline-alkali (64). Saline soils have an ECe of greater than 4 mmho/cm and an exchangeable sodium percentage (ESP) of less than 15 percent. Nonsaline alkali soils have an ECe of less than 4 mmho and an ESP of greater than 15 percent. Saline-alkali soils have both high salts (ECe of greater than 4 mmho/cm) and high sodium (ESP more than 15 percent). "Normal" soils have both low salt and sodium concentrations.

Saline soils have a sufficient concentration of salts to impair crop growth (5, 64). The soil is usually flocculated because of the high-salt concentration and drainage is not limited due to the salts. There may be white crusts on the soil surface, and there may be free carbonates or gypsum in the soil. With sufficient water passing through the soil and assuming the drainage water is removed from the root zone, the excess salts can be leached.

Nonsaline alkali soils have a high pH (8.5–10) with high sodium causing dispersion of soil colloids. This is caused by the incomplete saturation of cation-exchange sites and swelling that occurs with high sodium concentrations. The higher the ESP, the greater the amount of dispersion of colloids. Soil-organic matter is dispersed as well and tends to concentrate on the soil surface, upon evaporation of water. This results in black spots; thus the name, black alkali soils. Clay minerals, being dispersed, are susceptible to movement in the soil and sometimes move downward, accumulating in lower horizons and leading to low permeability. With the rising pH, calcium and magnesium tend to precipitate, thus increasing the ESP and associated problems.

When both salt levels and ESP are high, saline-alkali soils may appear similar to saline soils. The pH is usually 8.5 or below because of the presence of high concentrations of other cations in addition to sodium. Unfortunately, leaching of these soils may remove the salts responsible for flocculation and also increase ESP value. This could create an alkali soil and greatly reduced water movement.

Management of salt-affected soils adds a further dimension to turf production. Wise use of irrigation water requires a good knowledge of drainage.

Drainage systems. Surface drainage can be an important means of water removal from production areas on many sites (14). Clearly, surface drainage is used effectively for managing many areas (2, 45, 67). In dealing with salt-affected soils, surface drainage may be beneficial only as a means of diverting saline water runoff from other areas away from the turf site. Surface drainage techniques include: 1) open ditches for larger volume drainage, 2) land forming, 3) bedding, 4) drainage terraces, and 5) field ditches (14, 60). Use of sprinkler irrigation when the soil is only wet a few inches deep on turf in arid and semiarid sites may tend to accumulate salts deposited by the water without surface removal.

Water management is the key to salinity control in crop production where salinity is a problem, or potentially so (45, 65, 67). Irrigation water quality, irrigation practices, and drainage are intricately involved in water management in salt-affected areas. Subsurface drainage is essential for removing excess salts existing in the soil profile.

Subsurface drainage can occur naturally on some soils that are uniform and have a deep water table. Careful investigation of soils through the use of Soil Survey Reports and on-site investigation by a drainage engineer will reveal whether such natural drainage is adequate for crop production needs. One should be aware of the potential for a high water table to develop when land is irrigated (17). Even where the water table is deep, irrigation with water in excess of crop needs, particularly with leaching of salts in mind, can cause a gradual rise of the water table to where it can interfere with crops. Salts contained in the water can be carried to the soil surface through capillary rise caused by evapotranspiration. This will potentially increase the salinity hazard. The quality of water in the water table is a crucial factor for potentially increasing salts in the root zone.

Subsurface drainage can be accomplished by installing tile or flexible tube drainage or with mole drains. Mole drains are a temporary means of drainage used sparingly in the U.S. Their low initial cost is an advantage, but their effects tend to be short lived. The practicality of mole drains is highly dependent on soil conditions (60).

Tile-and-tube drainage. Use of tile (or tube drainage) for subsurface drainage provides the most effective and uniform means of water removal in areas where surface drainage is inadequate. Tile and/or tubes are installed in the bottom of a trench at the appropriate depth and grade so fill water can be drained (14, 60). The trench is filled back in and normal plant production follows.

For many years tiles were constructed of clay or concrete, requiring careful and tedious installation (19). Plastic-drain tubes were researched from 1947 to 1955 (59). Several other materials have been tried, but plastic tubing now constitutes a major part of the market because of its light weight, moderate cost, and ease in handling and installation (19).

In evaluating the kind of material to utilize, consider that the tile should be: 1) strong enough to support both static (soil weight) and impact (farm equipment) loads, 2) resistant to weathering, 3) resistant to water absorption, 4) resistant to damage by freezing and thawing, and by soil chemicals, including soluble salts, 5) uniform in size, shape, and wall thickness, and 6) free of defects (58).

Climate, soil, and crop must be considered when determining drainage needs. Depth of tile placement is important in using drainage for salinity control. In humid regions tile depth is often 2 to 4 feet. In irrigated arid areas tile are usually placed deeper (6 feet or more) to prevent capillary rise of saline water from the water table (45, 46). The proper depth of tile placement is related to spacing of tile lines. If more intensive drainage is desired, tile lines can be placed closer together and shallower. If the primary objective of the tile system is for salinity control, tile will be placed deeper and closer together. Speed of water removal is not as important as a gradual leaching of salts. Keep in mind: Spacing and depth should be such that the draw-down curve will keep the water table deep enough so that capillary rise will not occur significantly at the point which is between the tile lines. Proper attention to design of the system should consider this potential.

In some cases, turf managers modify the soil to improve drainage characteristics (2, 45). Most notable is the use of high-sand content soils in construction of golf greens. This allows for internal drainage as well as for easy leaching of salts. Installation of a gravel layer beneath the topsoil mix causes a perched water table at the interface and provides a natural barrier to upward capillary movement of saline water from a water table that might exist beneath the green. However, use of a salt-laden water for irrigating turfgrass grown above a perched water table or nonpermeable layer can cause salt accumulation and little recourse to leaching for salt removal.

Cultivation of turfed soils can improve infiltration (2, 45). For sites where salinity is a problem, particularly on sloping areas, core aeration will improve infiltration and the potential for leaching of soluble salts from the root zone.

Leaching of salts

When managing a salt-affected soil one works toward maintaining a salt balance in the soil, moving as much salt out of the root zone as is being added (57). This assumes that the salinity level in the soil is not high enough to significantly affect crop growth. Based on earlier work (64), it has been proposed (56) that the change in salt balance (ΔS_{sw}) is a function of inputs and outputs to soil water salinity according to the equation:

$$\Delta S_{sw} = V_{iw}C_{iw} + V_{gw}C_{gw} + S_m + S_f - V_{dw}C_{dw} - S_p - S_c$$

where V_{iw}, V_{gw}, V_{dw} and C_{iw}, C_{gw}, C_{dw} are volume and total salt concentration of irrigation, ground, and drainage waters, respectively. V_{gw} refers to that water which moves into the root zone from the water table; S_m is salt brought in to solution from weathering of soil minerals and dissolving salt deposits; S_f is the salts added from fertilizers, amendments, and manures; S_p is the salt that precipitates out in the soil from added irrigation water; and S_c is the salt removed by the harvested portion of the crop. If we assume that V_{gw}, S_m, S_f, S_p and S_c are zero, the equation reduces to $\Delta S_{sw} = V_{iw}C_{iw} - V_{dw}C_{dw}$. Further, if we assume a steady state condition where ΔS_{sw} is zero, then we have: $V_{iw}C_{iw} = V_{dw}C_{dw}$. This can be changed to $\frac{D_{dw}}{D_{iw}} = \frac{EC_{iw}}{EC_{dw}}$ where D is equivalent to the depth of water, substituting for volume, and EC is the electrical conductivity of the water, a measure of concentration. With these assumptions there is a direct relationship between the leaching fraction ($LF = D_{dw}/D_{iw}$) and the ratio of the salinity of the irrigation and drainage water (EC_{iw}/EC_{dw}). By varying that fraction of the applied water that percolates through the root zone, it is possible to control to some degree the salt concentration in the drainage water and, as a result, also control to some degree the salinity level of the soil water in the root zone. Lysimeter studies have been used to prove the validity of this approach (8). One researcher (56) concluded from these and other studies: 1) the salt content of soil water increases with depth in the root zone, unless irrigation occurs with low salt water and very high leaching fractions; 2) soil-water salinity for irrigation waters of a given salt content is essentially uniform near the soil surface, regardless of leaching fraction, but increases with depth as leaching fraction decreases; 3) average rootzone soil-water salinity increases and crop yield decreases as irrigation water salinity increases and leaching fraction decreases; and 4) the first increments of leaching are the most effective in preventing salt accumulation in the soil water of the root zone. The amount of water removed from the root zone by the crop will also affect the salinity level in the root zone.

An example of the application of the leaching requirement equation would be to assume a crop where a value of EC_{dw} of 8 mmho/cm can be tolerated and the irrigation water has EC_{iw} of 2. The leaching requirement will be 2 over 8 or 25 percent. This is a maximum value because of the assumptions made. Any changes in the system, such as leaching caused by rainfall, will negate application of this equation.

There is an increasing awareness that leaching of salts from production fields should be limited to protect drainage water from continuously increasing in salt content. River water used for irrigation and leaching of salts can move salt through the drainage system back into the river. Thus, the salt load increases as the water moves downstream. To offset this, use of lower leaching fractions has been studied. Reportedly, with use of less water for leaching there is more precipitation of calcium salts, particularly calcium carbonate and gypsum (33, 34). Under these conditions, the water uptake by the plant occurs primarily in the upper part of the root zone, and the salt content of the lower portion of the root zone increases considerably because of less leaching. At some point in time leaching of this increased salt load must take place for future cropping.

Several studies of the best means of water application to cause effective leaching have been made. Most early work was based on ponding with a one-time application. Others (48, 61) have shown that intermittent ponding was more efficient than continuous ponding to achieve the same degree of leaching, and application for leaching by sprinkling was more efficient than the other methods (49). Lighter and more frequent irrigations are most effective, especially on lower permeability soils (46, 61).

Water quality is an important factor in leaching requirements. Use of saline irrigation water will result in less than optimum plant growth (8).

It is difficult to predict a precise leaching fraction to leach the salts effectively and efficiently because soil varies and irrigation water is not applied uniformly. Field studies have shown predicted leaching fractions to be highly variable for different crops and soils (39).

Saline soils cannot be reclaimed by chemical amendment, conditioner, or fertilizer. Leaching can remove salts from the root zone. The amount of water required for leaching relates to the initial salt level in the soil, the final level desired, and the quality of the irrigation water (18).

In a simplification of leaching requirements for saline soil

6 acre-inches per acre of good quality leaching water passing through a foot of soil will reduce salinity by about 50 percent (18). One acre-foot per acre will reduce salinity by about 80 percent and 2 acre-feet per acre passing through 1 foot of soil will reduce salinity by about 90 percent.

Chemical treatment of sodic soils. Soils with high sodium levels require special treatment so that they can be used effectively in agricultural production. Upon leaching of saline-sodic soils, the loss of salts (electrolytes) may leave the soil with a higher ESP, possibly creating sodic soil. Materials suggested for treatment of sodic soils include calcium chloride, gypsum, sulfur, sulfuric acid, iron and aluminum sulfates, and limestone (64). Calcium chloride, while very quick and effective, is too costly and is not widely used (64). Use of sulfuric acid is hazardous to people and plants and causes corrosion of equipment. Sulfates of iron and aluminum are too expensive for practical application. Ground limestone is very effective on acid soils, but it is of less value in high pH soils. Thus, gypsum and sulfur are the two materials with greatest potential for use in reclaiming sodic soils.

The reaction of gypsum in sodic soils is:

$$2NaX + CaSO_4 \rightarrow CaX_2 + Na_2SO_4$$

where X represents the cation exchange site on the soil colloid.

Sodium is replaced on the cation exchange site of the clay mineral by calcium, forming sodium sulfate in the soil solution. Because of its low solubility, adequate irrigation water must be applied to dissolve the gypsum. It has been estimated that 12.5 metric tons (2,205 pounds) of gypsum/ha (2.47 acres) will be required to replace 1 mg exchangeable sodium/100 g soil (56). Further, it is estimated that 1 meter (3.3 feet) of irrigation water will dissolve 7.3 tons of gypsum/ha. The total amount of irrigation water needed to dissolve the 12.5 metric tons of gypsum/ha is 1.7 meters, replacing 1 mg sodium/100 g soil. Several leachings will be needed to accomplish the dissolution of the gypsum and exchanging and leaching of the sodium.

General recommendations for gypsum application are 45 metric tons gypsum per ha the first year with about 1.5 meters of water for leaching (56). In reclaiming a sodic soil, one could then begin by growing a shallow-rooted crop. Annual applications of gypsum and irrigation water for leaching must follow. In 4 or 5 years reclamation may be complete, depending on soil, water, and crop conditions. The objective is to get the ESP below 10 percent on fine-textured soils and below 20 percent on coarser-textured soils (5).

Sulfur can also be used to reclaim sodic soils. Sulfur must be oxidized by soil organisms; thus, the reaction will be slow. Soils must be warm for the sulfur-oxidizing organisms to be active. Reactions for sulfur in soils with free lime present are:

$$2S + 3O_2 + 2H_2O \rightarrow 2H_2SO_4$$
$$CaCO_3 + H_2SO_4 \rightarrow CaSO_4 + H_2O + CO_2 \uparrow$$
$$2NaX + CaSO_4 \rightleftarrows CaX + Na_2SO_4$$

In soils without free lime the reaction is:

$$2S + 3O_2 + 2H_2O \rightarrow 2H_2SO_4$$
$$2NaX + H_2SO_4 \rightleftarrows 2HX + Na_2SO_4$$

In acid soils the use of limestone gives the following reaction:

$$2HX + CaCO_3 \rightleftarrows CaX + H_2 + CO_2$$

Improvement of the physical condition of soil involves more than replacing sodium on cation-exchange sites. Excess sodium causes dispersion of the soil colloids so that they act independently instead of serving as a part of soil aggregates. Once dispersed, the soil colloids can move in the soil and accumulate in lower horizons, causing restricted drainage (47). Tillage of a dispersed soil magnifies this problem. To encourage the redevelopment of good soil structure, one must pay careful attention to good soil management practices—tillage at the proper moisture content, returning ample portions of organic material to the soil to encourage organism activity, growing healthy crops with good root systems, maintaining a good drainage system, and careful irrigation with attention to water quality. Use of salt- or sodium-tolerant crops may be necessary until reclamation is complete.

Several authors have suggested use of saline water to reclaim sodic soils (33, 34, 35, 54). The presence of excess salts results in maintaining flocculated soils in the presence of considerable sodium. An occasional leaching is necessary to avoid excess salt accumulation. Careful use of such waters can actually decrease the amount of salt leached in drainage water.

Blending water

Improvement of poor-quality waters can be accomplished by diluting them with better water. Frequently, a well with relatively high salts can be used as a water source if better quality water is available. The two water sources can be pumped into a reservoir with mixing allowed; then used to irrigate. The amount of salted water that can be used depends on the quality of the two water sources, climatic conditions, the soil, and the type of turf grown.

Salt-tolerant grasses

Excess soil salinity can prevent or delay seed germination; also, it may significantly influence plant growth. The first sign of salt influence, as with drought, is the blue-green appearance of a turf. There may be tip dieback, followed by thinning and death of the turf. Usually, salinity problems can be associated with heat and drought stress.

A recent summary of the effects of salinity on plant growth observed that it was once believed that plant growth reduction caused by salinity simply resulted when water uptake was reduced because of a decrease in the gradient between root or leaf cell solute concentration and salinity of the root medium (21). An increasing amount of experimental data and other evidence indicates that this "water availability" theory is invalid or, at least, a vast oversimplification of the true situation. In the field, salt uptake and accumulation provide a better basis for explaining growth phenomena that occur under conditions of excess salinity. Visible plant injury, growth suppression, and other internal changes in the plant are more directly related to increased internal ion (salt) concentrations and to ionic proportions.

As might be expected, grasses vary widely in the ability to

tolerate soluble salts. It is a common practice to plant tolerant grasses where salt problems are anticipated. However, use of grasses known to have a high salt tolerance may only be a short-term solution (9). And, copious irrigation and leaching can reduce salt levels and change a grass population from one with high salt tolerance to one of moderate-to-low tolerance. Consequently, whether leaching occurs naturally or through irrigation and drainage improvement, the more permanent solutions provided in this fashion need to be weighed against, or along with the use of salt-tolerant grasses.

Research in California in the 1960s pointed out the high salt tolerance of certain grasses (42). The salinity tolerance of 'weeping alkaligrass' was very high (it survived relatively well at 330 meq./sol. salts/l. of substrates). Following weeping alkaligrass in order of decreasing tolerance were 'Seaside' creeping bentgrass and 'Alta' tall fescue; Kentucky bluegrass and 'highland' bentgrass (*Agrostis tenuis*) were the least tolerant of the five grasses studied. They survived relatively well at 74 meq. Subsequent work (43) in California indicated good tolerance and only moderate growth reduction of weeping alkaligrass and Seaside bentgrass at ESP values of 26 to 28.

Alta tall fescue, Kentucky bluegrass, and common bermudagrass had growth reductions of about one-third to one-half in this range. Weeping alkaligrass has been reported growing, under golf course conditions, at up to 14 mmhos/cm in Colorado (10) and along roadsides in Illinois at levels of approximately 4,000 to 10,000 ppm Na and 20,000 to 30,000 ppm total soluble salts (9). There has been adequate research and observations to demonstrate that alkaligrass will tolerate both high levels of Na and total soluble salts in the soil and in irrigation water (25, 26, 27). In fact, Lemmon and weeping alkaligrass can apparently be germinated and grown in soil with good drainage using water as high in salt as 50 percent seawater, as long as enough moisture is provided to prevent drought (25).

Because of the ability of alkaligrass to tolerate high salt levels, it may develop into a noticeable weed problem where salts restrict growth of less-tolerant turfgrasses. General salt tolerance rankings of several grasses and forbs have been made (2, 3, 62, 64). Table 5 from USDA Agricultural Bulletin 194 (3) lists several grasses and forbs used primarily for forage and revegetation projects.

Table 6 gives a generalized listing of grasses, based upon observation and research as to salt tolerance, that can be used for turf in Colorado (62).

Information on salt tolerance of grasses has mostly been on species differences. A 1967 study found a wide variability in salinity tolerance among creeping bentgrass cultivars (73). In this work 'Arlington', 'Seaside', 'Pennlu', and 'Old Orchard' were found to be most tolerant; 'Congressional' and 'Cohansey' were found intermediate in tolerance, and 'Penncross' was the least tolerant. There were indications that the possibilities for developing strains with greater salinity tolerance are promising. On the other hand, a comparison of many cultivars of Kentucky bluegrass did not show an extensive range of salt tolerance (1). Of the cultivars tested, Nugget was the most tolerant. Also, 'Dawson' and 'Golfrood' fine fescue (*Festuca rubra* var. *trichophylla*) were found to be more salt tolerant than other *Festuca* spp., 'Ruby', 'Ranier', 'Steinacher', and 'Illahee'. The third-ranked group included 'Scaldis', 'Centurion', 'Pennlawn', and 'Atlanta', followed by the least-salt-tolerant group of 'Winter-

TABLE 5 Salt tolerances of grasses and forbs (3)

Good salt tolerance
(12 to 6 millimhos)

Alkali sacaton (*Sporobolus airoides*)
Saltgrass (*Distichlis stricta*)
Nuttall alkaligrass (*Puccinellia nuttalliana*)
Bermudagrass (*Cynodon dactylon*)
Tall wheatgrass (*Agropyron elongatum*)
Rhodes grass (*Chloris gayana*)
Rescue grass (*Bromus catharticus*)
Canada wildrye (*Elymus canadensis*)
Western wheatgrass (*Agropyron smithii*)
Tall fescue (*Festuca arundinacea*)
Barley (hay) (*Hordeum vulgare*)
Birdsfoot trefoil (*Lotus corniculatus*)

Moderate salt tolerance
(6 to 3 millimhos)

White sweetclover (*Melilotus alba*)
Yellow sweetclover (*Melilotus officinalis*)
Perennial ryegrass (*Lolium perenne*)
Mountain brome (*Bromus marginatus*)
Hardinggrass (*Phalaris tuberosa* var. *stenoptera*)
Beardless wildrye (*Elymus triticoides*)
Strawberry clover (*Trifolium fragiferum*)
Dallisgrass (*Paspalum dilatatum*)
Sudangrass (*Sorghum sudanese*)
Hubam clover (*Melilotus alba* var. *annua*)
Alfalfa (*Medicago sativa*)
Rye (hay) (*Secale cereale*)
Wheat (hay) (*Triticum aestivum*)
Oats (hay) (*Avena sativa*)
Orchardgrass (*Dactylis glomerata*)
Blue grama (*Bouteloua gracilis*)
Meadow fescue (*Festuca elatior*)
Reed canary (*Phalaris arundinacea*)
Big trefoil (*Lotus uliginosus*)
Smooth brome (*Bromus inermis*)
Tall meadow oatgrass (*Arrhenatherum elatius*)
Milkvetch (*Astragalus* species)
Sourclover (*Melilotus indica*)

Poor salt tolerance
(3 to 2 millimhos)

White Dutch clover (*Trifolium repens*)
Meadow foxtail (*Alopecurus pratensis*)
Alsike clover (*Trifolium hybridum*)
Red clover (*Trifolium pratense*)
Ladino clover (*Trifolium repens* forma *giganteum*)
Burnet (*Sanguisorba minor*)

green', 'Waldorf', 'Jamestown', 'Firmaula', 'Barok', and 'Durar', which were all dead at the conclusion of the study.

In work to determine the relative salinity resistance of 31 Kentucky bluegrass cultivars, it was found that nine cultivars resisted less than 1.0 mmhos/cm ECe and four ('Delta', 'Park', 'Nugget', and 'Prato') cultivars had a resistance of 20 to 23 mmhos/cm (30). The authors did caution that information gained from work with solution culture may not relate to field conditions.

Research on salt tolerance of turfgrass species and cultivars at the University of Nebraska (36) used .8 percent NaCl and $CaCl_2$ solutions. In the study six grasses were studied. 'Fults' alkaligrass was the most tolerant and 'Adelphi' Kentucky blue-

TABLE 6 Selected grasses with approximate salt tolerance (62)

Kind of grass	Salt tolerance* ECe mmho/cm	Grass type
For fine, domestic areas		
Kentucky bluegrass	3–4	sod-former
Perennial ryegrass	8–10	bunch
Red fescue	8–12	sod-former
Bermudagrass	16–18	sod-former
Alkaligrass	20–30	bunch
For coarse, naturalized areas		
Blue grama	5–6	weak bunch
Smooth brome	6–8	sod-former
Orchardgrass	6–8	bunch
Tall fescue	8–10	bunch
Western wheatgrass	12–16	sod-former
Tall wheatgrass	12–16	tall bunch

*Levels at which noticeable plant growth reduction and management problems usually occur.

grass was the least salt tolerant. Nugget was significantly more tolerant that Adelphi, and 'K-31' tall fescue was more tolerant to NaCl than Nugget, but Nugget tolerated $CaCl_2$ better than 'K-31'. Other studies found no significant differences among four tall fescue cultivars; however, the tall fescue cultivars were more salt tolerant than meadow fescue (*Festuca elatior*) (36).

Information is limited on the salinity tolerance of warm-season turf. One observer (2) indicates that bermudagrass, Zoysiagrass (*Zoysia* sp.), and St. Augustinegrass (*Stenotaphrum secundatum*) have good salinity (ECe 8–16 mmhos/cm) tolerance. It is not unusual to see St. Augustinegrass growing directly into seawater. *Paspalum vaginatum* var. Adalayd and Futurf reportedly have a high tolerance to salinity (28). In one instance, both varieties did well in soils of EC_e of 25 mmho/cm. Elsewhere, Futurf grew in soils with EC_e of 40–45 mmho/cm. It has been reported (Table 7) that blue grama and buffalograss, native to the High Plains of the U.S., perform poorly when grown in salt solutions in pot culture (36).

Salt problems, because of variability in soils and irrigation, vary considerably in severity on a site. To alleviate the possibility of stand-establishment failure and to produce a higher-quality turf, mixtures of sod-forming cool-season grasses, such as Kentucky bluegrass and creeping red fescue, are often mixed with more salt-tolerant bunchgrasses, such as perennial ryegrass (*Lolium perenne*) and alkaligrass (62). Commercial mixtures of such turfgrass seed are available. Because sod is often placed on salty sites, it has been proposed that grasses with different salinity tolerance be grown for sodding such sites (9).

Literature cited

1. AHTI, K., A. MOUSTAFA, and H. KAERWER
   1980. Tolerance of turfgrass cultivars to salt. Proc. Third Int. Turfgrass Res. Conf. 165–171.
2. BEARD, J. B.
   1973. Turfgrass: Science and Culture. Prentice-Hall, Inc. Englewood Cliffs, N.J. 658 pp.
3. BERNSTEIN, LEON
   1958. Salt tolerance of grasses and forage legumes. USDA Agr. Bull. 194. 7 pp.
4. BERNSTEIN, LEON
   1967. Quantitative assessment of irrigation water quality. *In* Water Quality Criteria, Amer. Soc. for Test. and Mat. Spec. Tech. Publ. No. 416. 51–65.
5. ———
   1974. Crop growth and salinity. *In* Drainage for Agriculture, Jan van Schilfgaarde (ed.). Amer. Soc. of Agron. Monograph 17. 39–54.
6. BERRY, W. L.
   1974. The use of effluent water in your management program. Proc. of 1974 Turf and Landscape Hort. Inst. (Univ. of Calif.). 48–50.
7. BLOSSER, R. O., and E. L. OWENS
   1964. Irrigation and land disposal of pulpmill effluents. Water Sewage Works III (9):424–432.
8. BOWER, C. A., G. OGATA, and T. M. TUCKER
   1969. Rootzone salt profiles and alfalfa growth as influenced by irrigation water salinity and leaching fraction. Agron. J. 61:783–785.
9. BUTLER, J. D., T. D. HUGHES, G. D. SANKS, and P. E. CRAIG
   1971. Salt causes problems along Illinois highways. Ill. Res. 35(4):3–4.
10. BUTLER, J. D., J. L. FULTS, and G. D. SANKS
    1974. Review of grasses for saline and alkali areas. Proc. Second Int. Turfgrass Res. Conf. 551–556.
11. CAMP, T. R., and R. L. MESERVE
    1974. Water and its impurities. Dowden, Hutchinson and Ross, Inc. Stroudsburg, Pa. 384 pp.
12. CECIL, L. K.
    1967. Municipal waters from western rivers. *In* Water Quality Criteria, Amer. Soc. for Test. and Mat. Spec. Tech. Publ. No. 416. 5–12.
13. DONALDSON, D. R., R. S. AYERS, and K. Y. KAITA
    1979. Use of high boron sewage effluent on golf greens. Calif. Turfgrass Culture 29(1):1–2.
14. DONNAN, W. W., and G. O. SCHWAB
    1974. Current drainage methods in the U.S.A. *In* Drainage for Agriculture, Jan van Schilfgaarde (ed.). Amer. Soc. of Agron. Monograph 17. 93–114.
15. EATON, F. M.
    1935. Boron in soils and irrigation waters and its effects on plants. USDA Tech. Bul. 448. 132 pp.
16. ———
    1950. Significance of carbonates in irrigation water. Soil Sci. 69:123–133.
17. FIREMAN, MILTON
    1957. Land drainage in relation to soils and crops. II. Salinity and alkali

TABLE 7 Visual quality ratings* of turfgrass species and cultivars treated with .8% NaCl and .8% $CaCl_2$ (36)

Turfgrass	Days after first salt treatment					
	34		59		80	
	Na	Ca	Na	Ca	Na	Ca
Fults alkaligrass	7.9 ab†	8.5 a	7.4 a	7.5 a	8.0 a	8.0 a
Buffalograss	6.6 c	7.6 c	3.9 c	5.6 c	2.4 d	2.9 d
Blue grama	4.7 d	6.8 d	2.7 d	4.0 d	1.8 e	2.1 e
Adelphi	5.0 d	6.6 d	1.5 e	4.0 d	1.0 f	2.1 e
Nugget	8.1 a	8.1 b	4.8 b	6.4 b	2.8 c	5.0 b
K-31	7.5 b	7.5 c	4.7 b	6.0 bc	4.0 b	4.5 c

*9 = ideal, 1 = dead

†DMR α = .05 within columns

problems in relation to high water table soils. *In* Drainage of Agricultural Lands, James N. Luthin (ed.). Amer. Soc. of Agron. Monograph 7. 505–513.

18. FOLLETT, R. H., W.T. FRANKLIN, and R. D. HEIL
    1979. Salt-affected soils. Service in Action. Colo. St. Univ. Ext. Serv. no. 503. 2 pp.

19. FOUSS, J. L.
    1974. Drain tube materials and installation. *In* Drainage for Agriculture, Jan van Schilfgaarde (ed.). Amer. Soc. of Agron. Monograph 17. 147–177.

20. FRANKLIN, W. T.
    1975. Criteria for evaluation of water quality for turf and ornamental growth. Proc. 21st Rocky Mtn. Reg. Turf. Conf. 55–58.

21. ———
    1983. Salinity and soil water (Syllabus for Agronomy 666). Dept. of Agron., Colo. St. Univ., Fort Collins. Unnumbered.

22. GALLATIN, M. H., J. LUNIN, and A. R. BATCHELDER
    1962. Brackish water sources for irrigation along the eastern seaboard of the United States. USDA Prod. Res. Rept. No. 61. 28 pp.

23. GARDENER, W. R., and M. FIREMAN
    1958. Laboratory studies of evaporation from soil columns in the presence of a water table. Soil Sci. 85:244–249.

24. HARIVANDI, M. A.
    1982. The use of effluent water for turfgrass irrigation. Calif. Turf. Culture 32(344):1–4.

25. HARIVANDI, M. A., J. D. BUTLER, and P. N. SOLTANPOUR
    1982. Effects of sea water concentrations on germination and ion accumulation in alkaligrass (*Puccinellia* spp.). Commun. in Soil Sci. Plant Anal. 13(7):507–517.

26. ———
    1982. Salt influence on germination and seedling survival of six cool season turfgrass species. Commun. in Soil Sci. Plant Anal. 13(7):519–529.

27. ———
    1983. Effects of soluble salts on ion accumulation in *Puccinellia* spp. J. Plant Nutrition 6(3):255–266.

28. HENRY, J. M., V. A. GIBEAULT, V. B. YOUNGNER, and S. SPAULDING
    1979. *Paspalum vaginatum* 'Adalayd' and 'Futurf'. Calif. Turf. Culture 29(2):9–12.

29. HILTIBRAN, R. C., and A. J. TURGEON
    1976. Bentgrass response to aquatic herbicides in irrigation water. Proc. Ill. Turf Conf. 17:34–44.

30. HORST, G., and J. B. BEARD
    1977. Salinity in turf grounds maintenance. 12(4):66, 69, 71, 73, 109.

31. JOHNSON, G. V.
    1975. Sewage effluent and turfgrass. Proc. of 1975 Arizona Turf. Conf. 4–5.

32. ———
    1976. Golf course irrigation with sewage effluent. The Golf Superintendent 44(6):26, 29, 41.

33. JURY, W. A., H. FRENKEL, D. DEVITT, and L. H. STOLZY
    1978. Transient changes in the soil-water system from irrigation with saline water: II. Analysis of experimental data. Soil Sci. Soc. Amer. J. 42:585–590.

34. JURY, W. A., H. FRENKEL, and L. H. STOLZY
    1978. Transient changes in the soil-water system from irrigation with saline water: I. Theory. Soil Sci. Soc. Amer. J. 42:579–585.

35. JURY, W. A., W. M. JARRELL, and D. DEVITT
    1979. Reclamation of saline-sodic soils by leaching. Soil Sci. Soc. Amer. J. 43:1100–1106.

36. KINBACHER, E. J., R.C. SHEARMAN, T. P. RIORDAN, and D. E. VANDERKOLK
    1982. Salt tolerance of turfgrass species and cultivars. Turfgrass Research Summary 1981. Univ. of Nebr., Dept. of Hort. Prog. Rept. 82-1. 85–95.

37. KOVDA, V. A., G. VAN DEN BERG, and R. M. HAGAN, eds.
    1967. International source-book on irrigation and drainage of arid lands. FAO/UNESCO. 663 pp.

38. KRENKEL, P. A., and V. NOVOTNY
    1980. Water quality criteria. Academic Press. New York, N.Y. 671 pp.

39. LONKERN, W. E., C. F. EHLIG, and T. J. DONAVON
    1979. Salinity profiles and leaching fractions for slowly permeable irrigated field soils. Soil Sci. Soc. Amer. J. 43:287–289.

40. LUDWICK, A. E., G. W. HERGERT, and W. T. FRANKLIN
    1976. Irrigation water quality criteria. Service-in-Action. Colo. St. Univ. Ext. Serv. no. 506.

41. LUNIN, J.
    1967. Water for supplemental irrigation. *In* Water Quality Criteria, Amer. Soc. for Test. and Mat. Spec. Tech. Publ. No. 416. 66–78.

42. LUNT, O. R., V. B. YOUNGNER, and J. J. OERTLI
    1961. Salinity tolerance of five turfgrass varieties. Agron. J. 53:247–249.

43. LUNT, O. R., C. KAEMPFFE, and V. B. YOUNGNER
    1964. Tolerance of five turfgrass species to soil alkali. Agron. J. 56:481–483.

44. MACKENZIE, A. J., and F. G. VIETS, JR.
    1974. Nutrients and other chemicals in agricultural drainage waters. *In* Drainage for Agriculture, Jan van Schilfgaarde (ed.). Amer. Soc. of Agron. Monograph 17. 489–511.

45. MADISON, J. H.
    1971. Principles of Turfgrass Culture. Van Nostrand Reinhold Co., N.Y. 420 pp.

46. MARSH, A. W.
    1969. Soil water irrigation and drainage. *In* Turfgrass Science, A. A. Hanson and F.V. Juska (eds.). Amer. Soc. of Agron. Monograph 14. 151–186.

47. MCNEAL, B. L.
    1974. Soil salts and their effects on water movement. *In* Drainage for Agriculture, Jan van Schilfgaarde (ed.). Amer. Soc. of Agron. Monograph 17. 409–431.

48. MILLER, R. J., J. W. BIGGAR, and D. R. NIELSEN
    1965. Chloride displacement in Panoche clay loam in relation to water movement and distribution. Water Resource. Res. 1:63–73.

49. NIELSEN, D. R., J. W. BIGGAR, and J. N. LUTHIN
    1966. Desalinzation of soils under controlled unsaturated flow conditions. 6th Congr. Int. Comm. on Irr. and Drainage. New Delhi. 19.15–19.24.

50. OERTLI, J. J., O. R. LUNT, and V. B. YOUNGNER
    1961. Boron toxicity in several turfgrass species. Agron. J. 53:262–265.

51. OSTER, J. D.
    1981. Salinity and its management. Water Management. Proc. 1981 Calif. Golf Course Supt. Inst. (Univ. of Calif.) 14–23.

52. RAINWATER, F. H.
    1962. Stream composition of the conterminous United States. U.S. Ged. Surv. Hydrol. Invest. Atlas HA–61. 3 pp.

53. RECORD, F. A., D. V. BUBENICK, and R. J. KINDYA
    1982. Acid rain information book. Noyes Data Corp. Park Ridge, N.J. 228 pp.

54. REEVE, R. C., and E. J. DOERING
    1966. The high salt water method for reclaiming sodic soils. Soil Sci. Soc. Amer. Proc. 30:498–504.

55. REEVE, R.C., and M. FIREMAN
    1967. Salt problems in relation to irrigation. *In* Irrigation of Agricultural Lands, R. M. Hagan, H. R. Haise, and T. W. Edminster (eds.). Amer. Soc. of Agron. Monograph 11. 988–1008.

56. RHOADES, J. D.
    1974. Drainage for salinity control. *In* Drainage for Agriculture, Jan van Schilfgaarde (ed.). Amer. Soc. of Agron. Monograph 17. 433–468.

57. RHOADES, J. D., R. D. INGVALSON, J. M. TUCKER, and M. CLARK
    1973. Salts in irrigation drainage waters: I. Effects of irrigation water composition, leaching fraction, and time of year on the salt compositions of irrigation drainage waters. Soil Sci. Soc. Amer. Proc. 37:770–774.

58. ROBERTSON, L. S., E. H. KIDDER, A. E. ERICKSON, and D. L. MOKMA
    1979. Tile drainage for improved crop production. Mich. Ext. Bul. E-909. Mich. State Univ.

59. SCHWAB, G. O.
    1955. Plastic tubing for subsurface drainage. Agr. Eng. 36:86–89, 92.
60. SCHWAB, G. O., P. W. MANSON, J. N. LUTHIN, R. C. REEVE, and T. W. EDMINSTER
    1957. Engineering aspects of land drainage. *In* Drainage of Agricultural Lands, J. N. Luthin (ed.). Amer. Soc. of Agron. Monograph 7. 287–394.
61. STEWART, B. A., and B. D. MEEK
    1977. Soluble salt considerations with waste applications. *In* Soils for Management of Organic Waste Waters, L. F. Elliott and F. J. Stevenson (eds.). Amer. Soc. of Agron. Ch. 9.
62. SWIFT, C. E., and J. D. BUTLER
    1982. Growing turf on salt-affected (alkali) sites. Service in Action. Colo. St. Univ. Ext. Serv. no. 7.227.
63. U.S. ENVIRONMENTAL PROTECTION AGENCY
    1977. The Report to Congress: Waste disposal practices and their effects on ground water. 512 pp.
64. U.S. SALINITY LABORATORY STAFF
    1954. Diagnosis and improvement of saline and alkali soils. L. A. Richards (ed.). USDA Handbook 60. 160 pp.
65. VAN SCHILFGAARDE, JAN,
    1974. Nonsteady flow to drains. *In* Drainage for Agriculture, Jan van Schilfgaarde (ed.). Amer. Soc. of Agron. Monograph 17. 245–270.
66. VIETS, JR., F. G., R. P. HUMBERT, and C. E. NELSON
    1967. Fertilizers in relation to irrigation practice. Irrigation of Agricultural Lands. R. M. Hagan, H. R. Haise, and T. W. Edminster (eds.). Amer. Soc. of Agron. Monograph 11. 1009–1023.
67. WADDINGTON, D. V.
    1969. Soil and soil-related problems. *In* Turfgrass Science, A. A. Hanson and F. V. Juska (eds.). Amer. Soc. of Agron. Monograph 14. 80–129.
68. WESSELING, JANS
    1974. Crop growth and wet soils. *In* Drainage for Agriculture, Jan van Schilfgaarde (ed.). Amer. Soc. of Agron. Monograph 17. 7–37.
69. WHITE, R. W.
    1980. The use of effluent in commercial sod production. Proc. 28th Ann. Fla. Turf-Grass Mgmt. Conf. 76–79.
70. WILCOX, L.V., G.Y. BLAIR, and C. A. BOYER
    1954. Effect of bicarbonate on suitability of water for irrigation. Soil Sci. 77:259–266.
71. WILCOX, L. V., and W. H. DURUM
    1967. Quality of irrigation water. *In* Irrigation of Agricultural Lands. R. M. Hagan, H. R. Haise and T. W. Edminster (ed.). Amer. Soc. of Agron. J. Monograph 11. 104–122.
72. YOUNGNER, V. B.
    1972. Irrigation of ornamental plants with effluent water. Proc. of 1972 Turf and Landscape Hort. Inst. (Univ. of Calif.). 87–89.
73. YOUNGNER, V. B., O. R. LUNT, and F. NUDGE
    1967. Salinity tolerance of seven varieties of creeping bentgrass, *Agrostis palustris* Huds. Agron. J. 59:335–336.

Practicum

8 Soil/water relationships in turfgrass

ROBERT N. CARROW

Robert N. Carrow received his B.S. and Ph.D. degrees from Michigan State University. He has held research and teaching positions in turfgrass at the University of Massachusetts and Kansas State University. Currently he is conducting research in turfgrass stress physiology and management at the University of Georgia at Griffin.

Turfgrass managers who want to conserve water must start with an understanding of the turfgrass plant, the soil, atmospheric effects, and water properties.

In the field these four components form a soil-plant-atmospheric continuum (SPAC) that influences water acquisition, movement, retention, and plant use.

The SPAC system can be presented in a budget form to delineate turf water management options. The budget approach is similar to a bank checking account with additions (inputs), withdrawals (outputs), and a reserve. The objective is to maximize inputs, minimize outputs, and maintain a large reserve.

Moisture inputs are precipitation, overhead irrigation, dew, and capillary rise of moisture from below the root system. Precipitation and overhead irrigation are the major inputs. Normally, capillary movement to turfgrass roots from below the root zone is minor, except where a water table is within 2 to 4 feet of the roots. On flat sod farms where the water table level can be controlled, capillary rise can contribute water for plant growth. Also, the PURR-WICK and USGA Green Section golf green construction methods use this principle. Drainage is impeded by a barrier (PURR-WICK) or perched water table (USGA) that results in more water being retained. This water can then provide some of the plant's water needs by capillary rise into the root zone.

Outputs or losses include runoff, leaching beyond the root zone, evaporation, and transpiration. Runoff would be a problem on sloped sites and can be increased by fine-textured soils, thatched turf, compacted soils, and application of water faster than the soil can receive it. Runoff causes not only a dry site but also an excessively moist site. Reducing runoff requires correcting these situations through cultivation, thatch control, or proper irrigation application rates.

Water movement beyond the root system is often an unrecognized water loss. Irrigators who water, based on the driest site, often overirrigate other areas. Irrigating *slightly* beyond the existing root system is acceptable because it provides a moist zone for further root extension. To reduce leaching losses, the irrigator must know the depth of rooting and depth of moisture penetration after applying a specific quantity of water. Well-designed irrigation systems that apply water uniformly will reduce leaching losses. Also, proper zoning of irrigation heads is important. Heads in similar areas should be zoned together. Poor zoning—with heads on slopes and low spots zoned together—results in poor uniformity.

Evaporation is the vaporization of water from a surface. In the SPAC system, evaporation is mainly from the soil surface and moist leaf surfaces. When moisture evaporates, it removes energy (heat) from the surface. Thus, evaporation helps cool the soil and plant if free water, such as dew, is on the leaf surface. Growers have considerable control over the quantity of water lost by evaporation. For example, evaporation from the soil surface is high immediately after irrigation, but it decreases dramatically as the soil surface dries. Thus, light, frequent irrigation results in high evaporative losses contrasted to losses after heavier, less frequent applications. Other ways to reduce evaporation are:

- Have good infiltration rates to get the moisture into the soil,
- Maintain a dense turf to shade the soil surface,
- Mow turf as high as feasible for your situation in order to ensure further shading,
- Avoid applying so much water that standing water occurs, and
- Avoid afternoon irrigation.

Transpiration is the vaporization of water inside the plant leaf, which then must diffuse through the open stomata. During this process, heat is removed from the plant. In many situations more than 90 percent of the moisture used by a turfgrass plant is used for cooling. Transpiration is a desirable use of water, especially in hot conditions. However, excessive transpiration wastes water. Overwatering promotes excessive transpiration.

The reserve of plant-available moisture at any point in time depends primarily on soil texture and the extent of the plant root system. Over a period of time, irrigation and precipitation are the sources of this reserve moisture. As a generalization, sands do not retain as much plant-available moisture as do loam soils. The turfgrass grower can markedly improve the moisture reserve by developing a deep, extensive plant root system. This will require careful mowing, irrigation, nitrogen fertilization, control of root-feeding insects, reduction of toxic substances (salts, herbicides) and, possibly, cultivation.

8. Soil/water relationships in turfgrass

Robert N. Carrow

The basics of soil

Soil and water are two fundamental resources necessary for turfgrass growth. Plant growth requires light, oxygen, carbon dioxide, water, mechanical support, and nutrients. Except for light, soil is involved in supplying these factors. A basic understanding of soil properties is essential for comprehending the complex interactions in the soil-plant-atmospheric continuum (SPAC).

Soil is a three-phase system composed of solids (soil matrix), liquid (soil solution), and gases (soil atmosphere) (fig. 1). Components of each phase are (a) solids—minerals and organic matter, (b) liquid—water plus dissolved substances, and (c) gases—carbon dioxide, oxygen, and other gases. While the soil matrix may change slowly over time, liquid and gas are dynamic and can change rapidly over time and space. Extremes in these phases, such as waterlogging (excess water) and dry soils (deficiency of water), markedly affect turfgrass growth.

Important characteristics of soils that influence plant growth are their chemical, biological, and physical properties. There are detailed reviews of soil properties (7, 18). Table 1 summarizes the most important chemical and biological properties. What follows is a brief overview of soil physical properties to provide the reader with a basic appreciation of the complex soil system. Soil physical properties include texture, structure, density, moisture, and aeration.

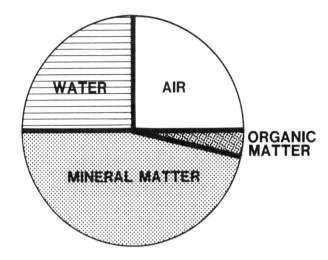

FIGURE 1. Illustration of the three phases of a typical loam soil. The approximate volume of each phase when the soil is at field capacity is a) solids 50 percent, composed of 47 percent mineral matter plus 3 percent organic matter; b) liquids 25 percent, and c) gases or air at 25 percent.

TABLE 1 Soil chemical and biological properties

Soil property	Comments
Chemical properties	
Soil colloids	Colloidal particles of clay and organic matter have very high surface area. Unless these particles are flocculated and aggregated, the soil will have poor aeration and drainage. Obviously a clay or organic soil has higher colloid content than a sandy soil.
Cation exchange capacity (CEC)	Colloidal particles have negative electrical charges which attract and retain cations (positively charged ions-K^+, Ca^{++}) from leaching. These cations are available for plant uptake. If the CEC has a high percentage of Na^+, the soil will become dispersed and have very poor physical properties (i.e., sodic soil).
Soil reaction	The soil mineral fraction contains many different chemical compounds which exhibit different solubilities depending on soil pH. Soil reaction (i.e., soil pH) refers to the hydrogen ion activity in the soil solution. A neutral soil has a pH of 7.0; an acid soil is <7.0, and an alkaline is >7.0. Soils generally range in reaction from pH 4.5 to 8.2. Depending on the pH, various essential nutrients become more or les available for plant uptake.
Biological properties	
Soil organisms	Soil contains many microorganisms—bacteria, fungi, algae, actinomycetes, and protozoa. These microorganisms are involved in many important processes such as: transformations of nitrogen, sulfur, phosphorus, and iron in soils; contribution to soil organic matter; and decomposition of soil organic matter. While most microorganisms are very beneficial, some may cause soilborne diseases. Macroorganisms (larger) also exist, such as nematodes, ants, moles, worms, and insects. These may at times cause problems on turfgrasses.
Soil organic matter	The transformations of organic matter in soils are complex. Basically, raw, undecomposed organic matter is continually broken down (decomposed) into more and more complex organic compounds. During this process microorganisms derive energy and food. The resulting material is termed humus, and because of its complex nature, humus is resistant to further decomposition. Humus aids in the formation of stable soil aggregates, which are essential for structure formation of heavy soils. Also, humus contains many chemical functional groups that contribute to soil CEC. During decomposition, many nutrients are made available for plant uptake.

Texture. The solid phase of soil does not change appreciably over time, except for soil formation, which requires hundreds of years. Solid particles consist of various primary minerals, secondary minerals (formed by reorganization of primary minerals after they have decomposed from weathering), amorphous mineral compounds (iron and aluminum oxides are examples), and humus. The amorphorus compounds and humus are often chemically and physically associated with the colloidal clay fraction.

A very important property of soil is soil texture: the size range of particles in the soil with three major textural fractions—sand, silt, and clay. The organic fraction is ignored unless soil organic matter content exceeds 20 percent (w) for sandy soils or 30 percent for fine-textured soils and such soils are more than 1 foot thick. If these criteria are exceeded, the soil is classed as an organic soil. All other soils are mineral soils. Names and size ranges of textural fractions are presented in Table 2 based on the U.S. Department of Agriculture classification.

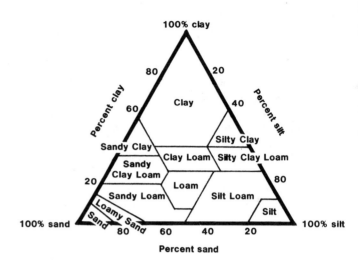

FIGURE 2. Textural triangle for classifying soils into textural classes based on the percent sand, silt, clay (wt. basis)

TABLE 2 Soil textural fractions

Separate	Diameter mm	Surface area in 1 gram (cm^2)	Type of mineral
Sand			Primary
Very coarse sand	2.00 – 1.00	11	
Coarse sand	1.00 – 0.50	23	
Medium sand	0.50 – 0.25	45	
Fine sand	0.25 – 0.10	91	
Very fine sand	0.10 – 0.05	227	
Silt	0.05 – 0.002	454	Primary
Clay	below 0.002*	8,000,000†	Primary/Secondary

*Colloidal clays are those smaller than 0.001 mm in diameter.
†From Sor and Kemper. Soil Sci. Soc. Am. Proc. 23:106, 1959.

Sand and silt fractions are relatively chemically inert since they do not have cation exchange capacity (Table 1). These fractions form the matrix or skeleton of soil and greatly influence moisture and aeration. The clay fraction does contain cation exchange capacity and therefore has a major influence on nutrient status. Clays can attract and hold water by adhesion and can fill in voids between the sand-silt matrix. As a result, clays also influence moisture and aeration.

The terms, light and heavy, are sometimes used and are misleading. A light soil is sandy in nature; a heavy soil is fine-textured. As we will see in the bulk-density section, these terms do not accurately reflect their weight.

By using weight ratios of the sand, silt, and clay fractions, soils can be assigned to one of 12 textural classes (fig. 2). For example, a soil with 40 percent (w) sand, 30 percent silt, and 30 percent clay would be in the textural class of clay loam. From knowledge about the textural class of a soil, information can be inferred about water-holding capacity, infiltration, aeration status for plants, and nutrient-retention capabilities.

Structure. Individual soil particles (sand, silt, clay) can be organized and arranged into what is called soil structure. Structure is especially important on soils containing appreciable silt and clay, since a good structure can open up the soil for better air movement, water drainage, and root growth.

Three broad categories of structure are recognized—single-grained, massive, and aggregated. A single-grained structure is typical of sands where the individual particles remain unattached to each other. A massive structure is demonstrated by a dispersed or compacted clay. The soil acts as one large mass. When soil particles are arranged into small, reasonably stable units (aggregates), the soil is termed aggregated. An aggregated soil is desirable.

Before an aggregate is formed, the soil must be flocculated. Flocculation is the physical joining together of clay particles to form larger units. Soils high in sodium can become deflocculated or dispersed by the large, hydrated sodium cation (Na$^+$) physically separating individual clay particles. Once flocculation occurs, the particles must be cemented together if a stable aggregate is to be produced. Cementation is provided by soil organic matter through various physical and chemical mechanisms. Some inorganic compounds, such as iron oxides, aluminum oxides, and calcium carbonate, also are cementing agents.

Aggregate formation can be the result of several forces, such as wetting-drying, freezing-thawing, root pressure, and mechanical cultivation. However, without stabilization by the cementation effects of organic matter, these aggregates will not be stable. Several forces destroy aggregates: raindrop impact, high-sodium levels in irrigation water, and traffic (i.e., soil compaction).

Bulk density and porosity. The density of a soil, called bulk density, is measured in mass of soil per unit soil volume. Bulk density reflects both solids and pore space. Pore space is the voids in soil that contain either air or water. The more dense a soil becomes, the higher the bulk density and the smaller the pore space. For example, compaction increases bulk density and decreases total pore space. Bulk densities generally vary from 1.1 g/cm^3 (well aggregated clay) to 1.7 g/cm^3 (fine sand).

Total pore space of soil is composed of many pore sizes. Large pores are important for root growth, air movement, and water drainage; small pores retain moisture. In a compacted soil, many larger pores are destroyed, while many small pores increase. The term aeration porosity is used to designate the proportion of soil volume filled with air at a given soil-water content or potential. Under field conditions it is desirable to have approximately 20 to 25 percent aeration porosity after a saturated soil has drained.

Soil air. As indicated, pore space is important because it influences soil air status. Plant roots and microorganisms use soil oxygen for respiration, while giving off carbon dioxide and other gases. Thus, soil atmosphere can differ in composition from aboveground atmosphere. Unless adequate gas exchange occurs, plant rooting can decline and microbial population changes can occur. The process of diffusion is the major means of gas exchange.

Factors that inhibit gas diffusion and thus cause poor soil aeration are waterlogging and soil compaction. Waterlogging may be the result of poor drainage, a high water table, or excessive application of water.

Soil temperature. Soil temperature is an important physical property of soil because it influences plant growth, microorganism activity, and, to some extent, water movement. The major source of heat in soils is radiation from the sun. High soil temperature at the soil surface is particularly detrimental to turfgrass root growth.

Surface soil temperatures, as well as temperatures deeper in the profile, can be moderated by a dense turf cover, some thatch, higher mowed turf, and a moist soil. In winter, snow, synthetic covers, and straw or pine needles reduce temperature extremes.

Soil water. The water molecule (H_2O) appears simple, but its properties greatly affect its movement and retention in soils. The H-O-H molecule of water has a bent configuration.

$$O^=$$
$$^+H \qquad H^+$$

While the whole molecule is electrically neutral, the arrangement of electrons in the molecule result in positive and negative regions (poles). Thus, a water molecule is called an electrical dipole. Because each water molecule has these poles, they are attracted to other water molecules (cohesion), electrically charged regions of soil particles (adhesion), or soil ions (adhesion). The weak mutual attraction between a hydrogen of one water molecule and the opposite charged oxygen of another molecule is called hydrogen bonding. (Note: Hydrogen bonding between water and another substance is adhesion.)

Surface tension is another property of water that results from its dipole nature. A water molecule below the water surface experiences equal cohesive forces on all sides (fig. 3). However, a water molecule at the surface experiences stronger forces into the denser water compared with the less dense gaseous phase.

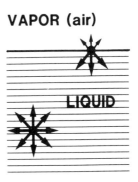

FIGURE 3. Illustration of surface tension in liquid water

The net result is a surface tension that acts like a thin film.

Essentially it is the combination of cohesion, adhesion, and surface tension that controls water movement and retention in soils. These properties also control water movement and retention in plants. Without these forces all water would simply drain from soils or a plant unless held by a physical barrier. Sometimes soil root-zone mixes are constructed to provide a physical barrier to water movement and thereby increase moisture retention. An example would be the PAT or PURR-WICK systems (1). In other cases root zone mixes have been constructed to take advantage of adhesion and cohesion; an example is the USGA method of greens construction (1).

Hydrogen bonds between adjacent molecules are much weaker than covalent bonds within a molecule. However, because there are many hydrogen bonds, the total bond energy (energy required to break a bond) is considerable. Within ice there is more hydrogen bonding than there is in liquid water and greater bonding in the liquid state compared with water vapor. Thus, to go from one physical state to another requires the input of energy (solid to liquid) or the release of energy (liquid to solid). When going from ice to liquid, 80 cal/g energy is required to disrupt the hydrogen bonds. When changing from the liquid to vapor phases, water absorbs 540 cal/g energy (latent heat of vaporization) to break the hydrogen bonds.

When a soil or a plant absorbs heat from its external environment, heat must be dissipated or temperatures will climb. Since it requires high energy to vaporize 1 g of water, this energy is removed from the plant or soil as the water vaporizes. Thus, this property of water allows temperature buffering and control in plant and soil systems through evaporation and transpiration.

With a basic appreciation of the properties of water, we can now look at soil-water relationships. After a soil is saturated, all the pore space is filled with water. Drainage starts due to gravity with the largest pores draining rapidly. After a few hours (1 to 2 hours for a sand, but 2 to 3 days for a heavy clay), drainage decreases to a very slow rate. This water content is often called field capacity. The concept of field capacity, used for many years, is not an exact characteristic of a soil (fig. 4). It is an approximation of the soil-water content after rapid drainage, but water redistribution continues even after the initial rapid phase. For irrigated turfgrass situations, where irrigation events are reasonably frequent, field capacity is a useful concept since water redistribution would be minimal. However, for infrequently irrigated or nonirrigated situations, it is wise to remember that the water

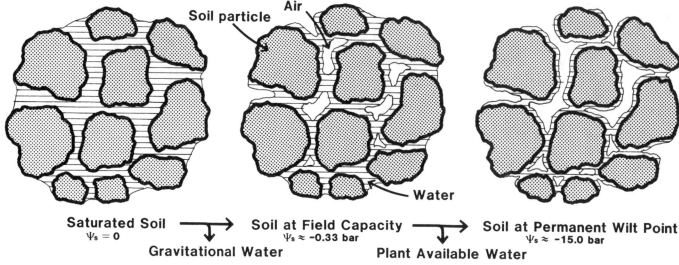

FIGURE 4. Field capacity—permanent wilt point concepts and their general relationship to soil water potential (Ψ_s)

content observed after rapid drainage will continue to decrease even if all evapotranspiration could be stopped.

Water held in the soil after rapid drainage is held by adhesion to soil particles and soil ions (solutes) as well as by cohesion. The first few layers of water molecules around a soil particle or solute are held rather firmly by adhesion, while the outer layers are held less tightly by cohesion.

Plants can easily extract moisture when the soil is near field capacity. As additional water is extracted, the remaining water is more difficult to obtain because it is held more firmly by the soil particles and solutes. Eventually, the soil water is held so tightly that plants can no longer extract enough for survival. The water content at this point is called the permanent wilt point (fig. 4). Permanent wilt point is not of particular use for turfgrass situations, since turf often exhibits wilting at water contents well above permanent wilt point.

TABLE 3 Total, plant available, and unavailable water-holding capacity of different soil texture classes

Soil texture	Water-holding capacity (inches per foot of soil)		
	Total	Available*	Unavailable†
Sand	0.6 – 1.8	0.4 – 1.0	0.2 – 0.8
Sandy loam	1.8 – 2.7	0.9 – 1.3	0.9 – 1.4
Loam	2.7 – 4.0	1.3 – 2.0	1.4 – 2.0
Silt loam	4.0 – 4.7	2.0 – 2.3	2.0 – 2.4
Clay loam	4.2 – 4.9	1.8 – 2.1	2.4 – 2.7
Clay	4.5 – 4.9	1.8 – 1.9	2.7 – 3.0

*Available for plant uptake.
†Not available for plant uptake.

The soil moisture held between field capacity and permanent wilt point is called plant-available water. Soil texture has a major influence on plant-available water with loams exhibiting the highest available water contents (Table 3). This value can be useful in determining how much moisture in the soil is available for plant extraction. For example, a well-structured silt-loam soil has the following characteristics: a total pore space of 55 percent by volume, which at saturation would be full of water; at field capacity a moisture content of 37 percent; and at permanent wilt point a content of 20 percent moisture.

$55\% \ H_2O_{vol} - 37\% \ H_2O_{vol} = 18\% \ H_2O_{vol}$
drained by gravity

$37\% \ H_2O_{vol} - 20\% \ H_2O_{vol} = 17\% \ H_2O_{vol}$
is plant-available water

Thus, in a 12-inch layer of this soil, the inches of plant-available moisture that would be present are:

(12-inch soil) (17%/100) = 2.04 inches plant-available water

In the next section, the concept of a water budget will be discussed and the importance to an irrigator in determining "plant-available moisture" in the turfgrass root zone will become evident.

For good-quality turfgrass, irrigation should be applied before the permanent wilt stage; thus, not all of this "plant-available water" will be truly available to the plant. However, the moisture that drains (18% H_2O_{vol} in this example) can be used by the plant until it percolates beyond the root system. Even with these offsetting factors, good-quality turf is often irrigated when between 25 to 50 percent of the plant-available moisture has been depleted (0.51 to 1.02 inches water in the example).

When a soil is moist, water films around particles are several layers thick and the outer layers are held by relatively weak forces. Upon drying, the water films become thinner with the inner layers held by stronger adhesive forces. Thus, not all soil water is equally available to plants. Recognition of this fact has led, in recent years, to a useful way to characterize soil water by its energy status or ability to do work. Soil water potential, ψ_s, denotes the energy status of soil water. Pure water is assigned a potential of 0 bar (bar is the most common unit to express ψ_s; 1 bar = 100 joules/kg = 100 centibar = 0.987 atm = 405.1 inches H_2O). Thus, soil-water potential is always 0 (saturation) or more negative. For example, soil moisture at field capacity usually has a ψ_s of between -0.1 to -0.33 bar; at permanent wilt point the ψ_s is -15.0 bar. Using the concept of energy status, one can now realize that as a soil dries, not only does water content decrease but the moisture that remains is less available (i.e., held more firmly) for plant uptake. Thus, soil water at $\psi_s = -5.0$ bar is less available for plant uptake than if at $\psi_s = -1.0$ bar.

Soil-water potential, ψ_s, is the sum of several components, where ψ_m is the matrix potential, ψ_o the osmotic (or solute) potential, and ψ_p is the pressure potential (fig. 5).

$$\psi_s = \psi_m + \psi_o + \psi_p$$

Matrix potential results from forces associated with the soil matrix-colloids, adsorption to soil mineral and organic matter, and cohesive water forces. Because these forces reduce water activity, ψ_m is always negative. It can be measured with tensiometers or in the lab with a pressure-plate apparatus.

Osmotic potential, ψ_o, is associated with solutes in soil. Solutes attract around them water films that reduce the activity of the water and thus its plant availability. Osmotic potential is also always negative.

Pressure potential, ψ_p, is the part of soil-water potential due to a pressure difference relative to the reference pressure. In unsaturated soils there is no liquid pressure, which results in $\psi_p = 0$. In some situations ψ_p may be positive, such as when a drainage barrier prevents normal drainage. The extra water retained in the soil is available for plant use. A drainage barrier

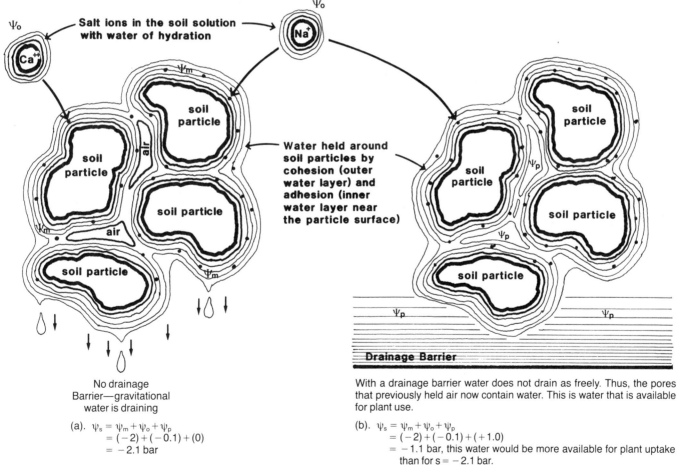

FIGURE 5. Concept of soil-water potential (Ψ_s) and its components. Ψ_m = matrix potential; water is less free due to cohesive and adhesive forces associated with the soil particles. Ψ_o = osmotic potential; water is less free due to cohesive and adhesive forces associated with the soil solutes or ions. Ψ_p = pressure potential; water is more available (or free) for plant use since it is not being retained by any forces attracting individual water molecules. $\Psi_s = \Psi_m + \Psi_o + \Psi_p$.

could be a polyethylene film, as in the PURR-WICK (1) system; a perched water table, such as in the USGA Green Section method of golf green construction (1), or a silt layer in the profile of a sand soil.

For most practical situations consider soil-water potential as the sum of the matrix and osmotic components, since for unsaturated conditions $\psi_p = 0$:

$$\psi_s = \psi_m + \psi_o$$

In most soils osmotic potential is very small since salt levels are minor. Exceptions would be salt-affected soils and soilless mixes in greenhouses. Thus, measuring ψ_m will provide a close estimation of ψ_s in most situations. Soil-water potential can be directly measured with vapor-pressure psychrometers.

Plant-water status can also be represented in energy terms. This aspect is discussed in detail in chapter 4, *Physiology of water use and water stress*, but a brief overview at this point may help the reader to understand the importance of water-energy status. Total plant- (or leaf-) water potential, ψ_L, is the sum of the osmotic (ψ_o), matrix (ψ_m), and turgor pressure (ψ_{TP}) potentials. Osmotic potential arises from reduced water potential due to dissolved solutes, while the matrix potential results from water attraction to the cell matrix, especially the cell wall. Both ψ_o and ψ_m are negative, since they reduce water activity. Matrix potential, generally considered to be very small, is often ignored.

Within a cell the ψ_o may be from -5 to -30 bar, which would cause water to flow into the cell from the intercellular space. As the cell expands, hydrostatic pressure (ψ_{TP}) develops and presses the cell membrane against the cell wall. This produces the turgidity observed on a plant that is not under moisture stress. Turgor pressure is positive since it tends to cause water to flow out of the cell.

The following situation illustrates the plant-water potential changes, as a plant goes from fully turgid (i.e., no moisture stress) to wilted:

Fully turgid $\psi_L = \psi_m + \psi_o + \psi_{TP}$
  $= 0 + (-10) + (+10) = 0$ bar

Wilted $\psi_L = 0 + (-14) + (+4) = -10$ bar

Note, as the plant wilts, ψ_o decreases due to a higher solute concentration, and ψ_{TP} decreases due to less pressure being exerted on the cell walls.

The ψ_L greatly influences water movement within the plant. Also, ψ_L affects many physiological (photosynthesis, nitrogen metabolism) and morphological (cell growth) processes in turfgrass plants.

Soil-plant-atmospheric system

While this chapter focuses on soil moisture, the influence of soil moisture on turfgrass growth can best be understood by seeing how soil moisture fits into the whole soil-plant-atmosphere system. Two useful ways of visualizing the interrelationships between soil-plant-atmosphere are the Soil-Plant-Atmospheric Continuum (SPAC) and the Budget Concept. The SPAC system is especially useful for researchers; it emphasizes water potential and the dynamics involved in water acquisition, movement, retention, and use. The Budget Concept is perhaps more useful to the turfgrass grower interested in efficient irrigation.

Soil-plant-atmospheric continuum. J. R. Phillip (14) first introduced the term SPAC for this dynamic and interrelated system. Anywhere in this system water has a particular energy potential, just as soil water can be characterized by its energy status. Typical ranges of water potentials are: soils, 0 (saturated) to -15 bar (dry); plants, 0 (fully turgid) to -20 bars (severely wilted), and atmosphere, 0 (raining) to -1000 bars (very low relative humidity). Water will move from a site of high-water potential (i.e., -1 bar) to one of lower potential (i.e., more negative, -5 bar) within this system.

Because a large water potential difference normally exists between soil and atmosphere, it appears that water can move easily from soil to plant to atmosphere. For flow to occur several resistances must be overcome: flow resistance of soil-water movement to roots; water movement across root cells to the xylem; water movement in the root; water movement in the stem, xylem, and leaf; and water movement from the leaf to the atmosphere, which requires a phase (liquid-to-vapor) change, and flow through the stomatal opening. The influence of such resistances can be readily observed on hot, dry days when turfgrasses exhibit wilt, even though soil-moisture content appears adequate. The turfgrass simply cannot extract or move sufficient moisture to meet its needs. The plant is not passive, however, and water stress can induce stomatal closure, thereby decreasing water loss.

The SPAC system is dynamic with all parts influencing plant-water status (fig. 6). A brief look at each component will

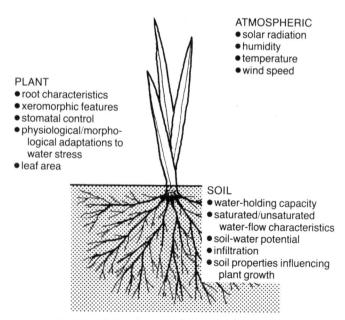

FIGURE 6. Soil-plant-atmospheric continuum (SPAC). These soil, plant, and atmospheric factors form a dynamic and interrelated system that influences water movement, plant uptake, and water potential.

illustrate this dynamic nature. The properties of soils that influence water movement and uptake are: soil-water potential, saturated and unsaturated flow characteristics, and soil-moisture content. Plant properties affecting water acquisition, movement, and retention are: rooting depth, rooting density, rate of growth, and xeromorphic features that reduce transpiration—location of stomata, thicker cuticles, decreased leaf area, rolling of leaf blades, and others (2). Also, plants may physiologically adapt to drought by allowing stomates to remain open at lower leaf-water potentials. Atmospheric conditions that stimulate rapid water uptake are: high light intensity, high temperatures, low humidity, and high wind movement. While these will stimulate transpiration, turfgrass plants may close their stomates under very high light intensity and temperatures which would drastically reduce transpiration.

To develop a good water management program, a turfgrass manager must essentially control as many as possible of the soil, plant, and atmospheric properties affecting water acquisition, retention, and use. This requires a high degree of knowledge of the dynamic SPAC system.

Budget concept of soil water. Another way to visualize turf water management is to consider a budget approach, similar to a bank checking account (fig. 7). Certain additions (inputs) of moisture are made and there are losses (outputs) of moisture from the plant environment. At any point in time, the plant has available to it a certain reserve of available water. The objectives of a wise turfgrass manager are to maximize inputs, minimize outputs, and maintain a large reserve.

Inputs of moisture are precipitation, overhead irrigation, dew, and capillary rise of moisture from below the root system. Precipitation and overhead irrigation are the major inputs. Normally, capillary movement to turfgrass roots from below the root zone is minor except where a water table is within 2 to 4 feet of the roots. On flat sod farms, where the water table level can be controlled, capillary rise can contribute water for plant growth.

Also, the PURR-WICK and USGA Green Section golf-green construction methods use this principle (1). Drainage is impeded by a barrier (PURR-WICK) or perched water table (USGA), which results in more water being retained. This water can then provide some of the plant's water needs by capillary rise into the root zone.

Outputs or losses include runoff, leaching beyond the root zone, evaporation, and transpiration. Runoff can be a problem on sloped sites and can be increased by fine-textured soils, thatched turf, compacted soils, and applying water faster than the soil can receive it. Runoff causes not only a dry site but also an excessively moist site. Reducing runoff requires correcting the above situations through cultivation, thatch control, or proper irrigation application rates.

Water movement beyond the root system is often an unrecognized water loss. Irrigators whose watering is based on the driest site often overirrigate other areas. Irrigating slightly beyond the existing root system is acceptable because it provides a moist zone for further root extension. To reduce leaching losses, the irrigator must know the depth of rooting and depth of moisture penetration after applying a specific quantity of water. Well-designed irrigation systems that apply water uniformly reduce leaching losses. Also, proper zoning of irrigation heads is important. Heads in similar areas should be zoned together. Poor zoning, with heads on slopes and low spots zoned together, results in poor uniformity.

Evaporation is the vaporization of water from a surface. In the SPAC system, evaporation is mainly from the soil surface and moist-leaf surfaces. When moisture evaporates, it removes energy (heat) from the surface. Thus, evaporation helps cool the soil and plant if free water, such as dew, is on the leaf surface. Excessive evaporation is wasteful. Growers can control the quantity of water lost by evaporation. For example, immediately after irrigation evaporation rates from the soil surface are high, but as the surface dries evaporation dramatically decreases. Thus, light, frequent irrigation results in high evaporative losses contrasted to heavier, less frequent applications. Other ways to reduce evaporation are: Have good infiltration rates to get the moisture into the soil; maintain a dense turf to shade the soil surface; mow your turf as high as feasible for your situation to insure further shading; avoid applying so much water that standing water occurs; and avoid afternoon irrigation.

Transpiration is the vaporization of water inside the plant leaf which then must diffuse through the open stomata. During this process, heat is removed from the plant. In many situations more than 90 percent of the moisture taken in by a turfgrass plant is utilized for cooling purposes. Transpiration is a desirable use of water, especially in hot conditions. However, excessive transpiration can occur and thereby waste water. Overwatering turf promotes excessive transpiration.

The reserve of plant-available moisture at any point in time depends primarily upon soil texture and extent of the plant-root system (fig. 7). Obviously, over a period of time, irrigation and precipitation are the sources of this reserve moisture. Soil-texture water-holding relations are detailed in the "water-holding capacity" section, but as a generalization sands do not retain as much plant-available moisture as do loam soils. The turfgrass grower can markedly improve the moisture reserve by developing a good deep, extensive plant-root system. This will require careful mowing, irrigation, nitrogen fertilization, control of

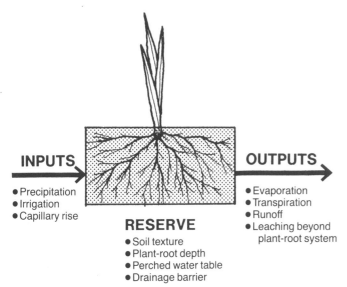

FIGURE 7. Budget Concept of turfgrass water management

root-feeding insects, reducing toxic substances (salts, herbicides), and possible cultivation.

With this background in soil-physical properties and the soil-plant-atmospheric continuum, we can now explore the major characteristics of soil that directly influence turfgrass water management. When reading this section, the reader should relate the information to the budget concept of water management. Also, Taylor and Ashcraft (15) present an excellent in-depth discussion of soil-water-plant relationships.

Soil characteristics affecting water management

Because plants obtain water from soil, it is with soil that water conservation must start. Certain characteristics of soil influence water movement and retention. This section's emphasis is on important soil physical problems that the turfgrass grower may confront.

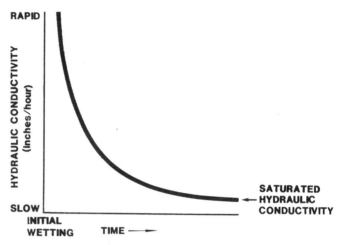

FIGURE 8. Infiltration rate over time following wetting a dry soil surface

Infiltration. Infiltration is the movement of water into the soil. Therefore, soil surface conditions are of prime interest, but other factors that influence infiltration are soil texture, structure, slope, and soil-water content. Sandy soils may have infiltration rates in excess of 1.0 inch per hour while clay soils are often less than 0.1 inch per hour (Table 4). A well-structured soil will provide large pores for infiltration, as contrasted to a tight clay or compacted soil. Organic matter often aids infiltration by providing a better-structured soil. Cracks that develop upon drying can markedly increase initial infiltration rates after irrigation. These are especially beneficial on heavy clay soils. A sloped surface will exhibit a much lower infiltration rate which complicates irrigation zoning and scheduling, particularly on rolling terrain such as golf courses. Thatch can reduce infiltration if it becomes hydrophobic (water-repelling) in nature. Turf managers have sometimes observed hydrophobic sands with very low infiltration rates. These are caused by organic coatings on the sand particles that repel water.

TABLE 4 Representative infiltration rates for different soil textures

Soil texture	Infiltration rate*	
	inches/hr.	cm/hr.
Sand—coarse	1.00 – 8.00	2.50 – 20.00
Sand—very fine	0.50 – 3.10	1.25 – 8.00
Sandy loam	0.40 – 2.60	1.00 – 6.50
Loam	0.08 – 1.00	0.20 – 2.50
Clay loam	0.04 – 0.60	0.10 – 1.50
Clay	0.01 – 0.10	0.02 – 0.25

*These values are approximate. Infiltration rates can vary widely, depending on surface conditions and water content.

When water is applied to a dry (or partially dry) soil, the initial infiltration rate is high and then gradually decreases. The final infiltration rate will depend on the water application rate. If the water is applied at equal to or above saturated hydraulic conductivity, the final infiltration rate will equal saturated hydraulic conductivity (defined in percolation section) (fig. 8). At water application rates below saturated conductivity, the steady-state infiltration rate will equal the unsaturated conductivity at the particular soil-moisture content.

The decrease in infiltration overtime is primarily due to a decrease in the matrix potential gradient. On a dry surface, ψ_m would be low (more negative), but as the surface few inches moisten the ψ_m approaches zero. Other factors influencing the drop in infiltration are the sealing of surface cracks, air entrapment ahead of the wetting front, and deterioration of surface structure.

Turfgrass growers have a choice of management techniques to improve infiltration rates. On soils with unfavorable infiltration rates, a good cultivation program can be very beneficial. Coring and slicing are effective; spiking may provide only short-term benefits. High traffic areas may require frequent cultivation—for example, coring once or twice per year and slicing every 2 to 3 weeks.

Allowing the soil surface to dry between irrigations can increase the initial infiltration rates by causing soil cracking and increasing the ψ_m gradient. On heavy clay soils, irrigators can program to apply a long burst of water initially, followed by shorter bursts with time in between for infiltration to occur. This would be in contrast to irrigating at very low rates (to match saturated infiltration rates) over long periods. Designers of irrigation systems should give attention to whether the particular soil has expanding clays that crack on drying. If cracks routinely occur, then systems with application rates higher than the saturated infiltration should be used.

Application of organic matter to the soil surface of poorly structured soils can improve infiltration. The easiest way to add organic matter is by growing a healthy plant and returning clippings. Well-digested sewage sludge, or other well-decomposed materials applied as a topdressing will be useful. However, these should be applied in conjunction with a good core aeration program to prevent surface layering.

Breaking up any surface layer, whether from compaction, or a distinct textural change, will improve infiltration. For example, sod containing a silt-loam soil laid over a sandy root zone mix will result in slow moisture penetration since the silt loam establishes the infiltration rate. When either partial, or complete modification of a root zone mix is attempted, careful attention should be given to the saturated infiltration rate. USGA Green Section golf green specifications require a rate of 4 to 6 inches per hour (1).

If a hydrophobic sand is the cause of poor infiltration, wetting agents may be beneficial to rewet the surface. Wetting agents can also enhance water penetration of a hydrophobic thatch. Sometimes irrigators use a short prewetting cycle to moisten a hydrophobic thatch. About 15 to 30 minutes after a light watering, the thatch will moisten sufficiently to allow heavier irrigation. If thatch is the main cause of low-water intake, dethatching is desirable.

Percolation. Percolation is the internal drainage of water in the soil. Water movement within the soil can either be by saturated flow (all soil pores filled with water) or unsaturated flow (at least some pores containing air). Darcy's law for steady-state (soil-water content remains constant) water flow in porous media is useful for understanding flow phenomena. This equation can be written as:

$$J = \frac{Q}{At} = K \frac{\Delta \Psi_s}{\Delta Z}$$

Where J = the quantity of water (Q) passing a unit section of soil (A) pores during a specific time period (t) (i.e., water flux density) in units of g cm^{-2} hr.

K = the hydraulic conductivity in cm hr^{-1}

$\Delta \Psi_s$ = difference in total water potential to the soil depth of Z(in cm)

$\Delta \Psi_s$ is the driving force for water flow with flow from a site of higher ψ_s to lower ψ_s.

Saturated flow is in the larger pores and the difference in pressure potential (ψ_p) is primarily responsible for the $\Delta \Psi_s$. As drainage continues, the larger pores are drained and unsaturated flow is initiated. Under unsaturated conditions, pressure potential is no longer important while the matrix potential (ψ_m) component to ψ_s becomes more important as a driving force for water flow.

A common observation is that hydraulic conductivity is not a static value but changes with soil-water content (fig. 8). Under steady state conditions, saturated hydraulic conductivity is greater than unsaturated conductivity. In fact, the conductivity at -15.0 bar may be only 10^{-3} as much as at saturation. The major factor that causes a reduction in hydraulic conductivity in unsaturated soils is the large decrease in cross-sectional area for liquid flow as the large pores drain. Thus, the drier a soil becomes, the less water redistribution occurs.

Saturated flow in sands is high, due to many large pores. In a clay soil, saturated flow is very low since few large pores are present. Thus, most water movement is through water films.

While saturated flow is primarily in the larger pores, unsaturated flow is water movement in the moisture films around and between soil particles. As a soil dries, the water films around the particles become thinner and less continuous, thereby reducing the flow rate. Soils with predominately fine pores will have low unsaturated flow rates because the pores offer high resistance to flow. At the other extreme of texture, unsaturated flow in sands is very slow because of few water films and fewer points of contact for film continuity. The direction of flow will be from a moist to a drier area. For example, a light rain can promote downward flow if the subsurface is drier than the surface; a water table 2 feet below the surface can result in upward movement (capillary rise) into the drier soil.

Internal drainage may need improvement on fine-textured soils to avoid excessive water problems such as poor aeration, soggy soils, intracellular freezing, and scald. Tile drainage can be used to remove subsurface water, providing an adequate outlet is available. Tile drainage is only effective on saturated soils. A wet, unsaturated soil will exhibit little water drainage into tile lines. If the water table is within 3 to 4 feet of the soil surface, the drainage can be useful. In humid climates, tiles are generally installed at 3 feet, but in arid climates installation at 5 to 6 feet may be necessary to avoid salination of soil surface. Upward movement of saline water to the surface, where it can evaporate, can increase salt levels at the surface. When deciding on whether tiling is necessary, the grower should evaluate the soil for impervious layers that impede internal drainage. Such a layer can produce a false water table that can be removed by physically breaking through the layer. Thus, tiling would not be required. The next section on layering discusses this situation.

Deep cultivation can increase internal drainage by allowing rapid water penetration to the depth of cultivation. A soil with a compacted surface of 1 to 2 inches—common for recreational sites—will respond to cultivation deep enough to penetrate completely the compacted zone.

Soil modification with physical amendments can improve the structure of fine-textured soils and increase their internal drainage. The addition of 5 to 15 percent by volume of a well-decomposed organic matter to clay soils is especially beneficial. However, internal drainage will be improved only to the depth of mixing. Also, the organic matter should be very well mixed into the soil. If surface compaction occurs, the site may still exhibit poor water infiltration, but cultivation could help correct this problem. Excessive application of organic matter should be avoided; it retains considerable moisture and produces a soggy soil.

Addition of sand to a fine-textured soil is sometimes attempted to improve internal drainage as well as infiltration. This has often been unsuccessful. Adding a solid sand particle to a clay soil essentially displaces the volume occupied by clay or silt. The same volume of clay or silt does contain considerable internal porosity composed primarily of small pores which, under moist conditions, retain water. Thus, the sand particle essentially eliminates what internal porosity has been present. It is not until sufficient sand particles are present that internal porosity (and drainage) starts to improve. This requires at least 80 percent by volume sand, and generally noticeable improvement does not occur until 85 to 95 percent sand. With these high percentages of sand, the sand particles start to contact or bridge with each other, thus opening up larger pores in the soil. Amending 1 inch of a clay soil requires a minimum of 4 to 8 inches of sand, uniformly mixed. The cost and logistics are often prohibitive. If sufficient sand is not added, the mix is often less desirable than the original. John Madison (10) provides a good discussion of this topic as well as other turfgrass soil problems.

Layered soils. Soil layering, a common occurrence on turfgrass sites, can be of two types: a) thin layers in the soil that differ in physical properties from the overlying or underlying soil and b) layered soils (thick layers) where the profile consists of distinctly different soils. An example would be 6 inches of sand over a silt loam soil.

Thin layers may arise from several sources. The most common causes: a) poor topdressing, where a topdressing mix contains excessive fine particles; b) sod containing organic matter or where silt is laid over a sand root zone mix; c) wind deposition of silt after a windstorm; d) water-deposited silt or clay; e) improper mixing of soil amendments, resulting in discontinuous layers; f) thatch, either on the soil surface or buried; and g) soil compaction at the surface 1 to 2 inches, which acts like a layering problem.

Layered soils may be due to natural soil formation and each layer is called a horizon. However, layered soils can be caused by human beings, who may, for example, bring in a topsoil distinctly different from the underlying soil. Such layers can be beneficial or detrimental, depending upon their characteristics.

Regardless of the type of layering, problems can arise in layered soils. Infiltration is reduced if a fine-textured layer occurs at the soil surface and it may be further restricted if soil compaction occurs. This explains why a sandy root zone mix can have poor infiltration if it is capped with a fine-textured layer.

Percolation, internal drainage, can be markedly reduced by even a thin zone of different texture. For example, a thin silt layer 6 inches below the surface of a sand soil can impede drainage and cause water to pond (temporary perched water table) above the layer because the hydraulic conductivity of the layer is less than that of the sand. Obviously, if the fine-textured layer is several inches thick, percolation can be even more affected.

Even a sandy layer in a silt or clay soil can disrupt water flow. While coarse-textured sand has a highly saturated conductivity, its unsaturated conductivity is low because of few water films and poor continuity of the films. Thus, water flow from above will continue to the sand zone and then start to accumulate until saturation above the layer occurs. After the saturated zone reaches a few millimeters, the pressure potential results in positive hydraulic pressure and free water (water not absorbed by the soil matrix or solutes) starts to drain. Once drainage starts, the sand layer will conduct the water rapidly.

Poor infiltration and percolation in turfgrass soils can result in a host of problems associated with excessive water. Examples: scald; intracellular freezing; encouragement of *Poa annua* L. infestation; disease—brown patch, pythium blight; poor cold, heat, or drought hardiness; poor aeration for root development; and interference with use and management of the turf because of wet soils.

As mentioned, poor aeration will reduce root growth, but layering has other influences on root development. A fine-textured zone at the point of root initiation and elongation may restrict growth due to mechanical resistance (only small pores for the roots to expand through) and low oxygen for respiration. Also, a root tip that comes into contact with a distinctly different texture zone may grow horizontally instead of penetrating more deeply. If root growth is restricted, turf-water management becomes more difficult.

In semiarid and arid regions, evaporation rates are high. As water evaporates from the soil surface, salts are left behind. Soils high in total salts or poor-quality irrigation water can promote salt deposition. Layers reduce percolation and therefore allow more capillary movement of water to the soil surface. In the section on water-holding capacity, capillary rise of water is detailed. The layers also inhibit leaching of salts from these soils. Whenever high salt levels accumulate near the surface, soil moisture becomes less available for plant uptake since the solutes in the soil solution reduce water activity.

Correction of layering problems depends upon its nature. Thin layers near the surface can often be corrected with core aeration. In layers composed of fine particles cores may be removed and holes filled with sand by topdressing. Deep-core aerators can be used to penetrate 6 to 8 inches. Also, subsurface cultivation or mechanically punching holes with pressurized water from a tree-root feeder can be effective.

Whenever soil of a distinctly different texture is applied over another soil, care should be taken to remove the sharp boundary difference. The underlying soil surface should be cultivated, some of the topsoil added and mixed into the surface 1 inch of underlying soil, and finally the remaining topsoil added. This will result in a gradual textural change.

In recent years sand topdressing has become popular on golf greens and sometimes on athletic fields. If the original soil mix is sandy, a sand topdressing should not result in a layering effect. However, with a fine-textured soil, the sand topdress should be in conjunction with core aeration and preferably core removal, followed by topdressing. This should be done for at least the first two to four topdressing applications.

Soil compaction. Effective water management can be greatly hindered by soil compaction. High-use recreational sites are especially susceptible. In most turfgrass situations compaction is confined to the surface 1 to 2 inches, but this narrow compacted zone can adversely affect plant growth and water movement. Soil compaction effects on turfgrass-water use are discussed in chapter 6, *Turfgrass culture and water use*. Here, let us look briefly at plant and soil characteristics altered by compaction.

Reduced total root growth has been observed by many researchers. O'Neil and Carrow (11) also observed an influence on root distribution with an increase in percentage of roots in the surface few inches but a dramatic decrease in deep-root growth. They also noted that compaction results in reduced aeration in the soil long after irrigation compared with a noncompacted site. This could reduce root viability and decrease water uptake by roots, if oxygen stress is sufficient to alter root permeability. The net effect of these plant-root responses is a less favorable condition for moisture uptake. Since the turfgrass exhibited decreased growth rates and total growth, this reduces water requirements and could partially offset decrease in water availability (11, 12).

Soil compaction reduces soil aeration and total porosity, while increasing bulk density, soil strength, and moisture retention, i.e., a shift to smaller pore sizes (5, 11, 12). These responses create an environment unfavorable for root penetration, water movement, and gas exchange. Infiltration and percolation rates can be greatly reduced, while soil oxygen levels become low. Compaction is most serious on soils high in silt or clay since these particles are pressed closer together into a more dense mass. High sand content soils resist compaction because the large sand particles provide a resistant matrix.

On compacted sites the turfgrass grower is confronted with many problems. Excess water problems, such as those mentioned in the layered soil section, are enhanced. The turfgrass becomes more susceptible to drought and high temperature stresses, due to limited rooting and a less dense turfgrass stand. Wear can be accentuated since shoot growth rates decrease under compaction (12). This fact is often overlooked by researchers and growers. One study (11) illustrated that after compaction was applied, shoot growth declined immediately while root-growth responses took several weeks before they were exhibited. Wear stress is greater under slow shoot growth. (Note: There has been much controversy over critical oxygen diffusion rates, ODR, for root growth of turfgrasses. Data indicate that perhaps oxygen becomes limiting for shoot growth before it does for rooting (12). A possible mechanism could be reduced root permeability at low oxygen, which could partially inhibit uptake of water. Without as much water, shoot growth would decline.) Other management problems would include difficulty in irrigation programming and increased maintenance costs.

Of four approaches to eliminating or reducing soil-compaction problems, one is to use species and cultivars that exhibit a greater tolerance to compaction. Unfortunately, present information is limited in this area. An evaluation of three cool-season grasses for compaction tolerance found perennial ryegrass ≥ Kentucky bluegrass > tall fescue (5). Several Europeans have evaluated traffic tolerance (wear-plus-compaction stresses). In many cases it appears that compaction is the primary stress in studies conducted on heavier soils. A general species tolerance from these studies (3, 4) is: annual bluegrass > perennial ryegrass > tall fescue > Kentucky bluegrass > red fescue > colonial bentgrass.

A second practice to reduce compaction is by spreading traffic around. Examples include using large tees and greens, using several football practice fields on rotation, moving tee markers and green flags, flairing the ends of cart paths, and using trees and shrubs to direct traffic. Irrigation timing versus traffic use can be important. The greatest degree of compaction occurs near field capacity. Thus, allowing an area to dry below field capacity before use can be beneficial. Also, avoiding light, frequent irrigation, which keeps soil surface moist, will reduce compaction.

A third cultural practice to reduce compaction is cultivation. If compaction is a problem, the turfgrass manager should develop a good cultivation program. This will also make irrigation easier. Coring and slicing are most effective.

The last alternative is soil modification, either partial or complete. As indicated previously, both well-decomposed organic matter and high sand content can effectively modify a fine-textured soil. In very high-use situations, complete modification may be the least costly over time. If the final root-zone mix is to resist compaction, it must have a very high sand content, in the 85 percent or more range.

Water-holding capacity. Soils differ in their total water-holding capacity as well as in plant-available water content (Table 3). Obviously, soil texture influences water-holding capacity. Table 3 reveals that loam soils have the most plant-available water. Sands have the fewest small pores for moisture retention. Clays have a high percentage of small pores, but moisture in many is retained too tightly for plant uptake. Sometimes sand soils are amended with soil containing more silt and clay to enhance moisture-holding capacity.

Organic matter can improve the water-holding capacities of sands. Often very high sand content soils or root zone mixes benefit from the addition of 5 to 15 percent by volume of a well-decomposed organic amendment. Too much organic matter may be detrimental by reducing aeration from excessive moisture retention in the organic matter.

Another factor that can increase soil-water content is control of the water table within 2 to 4 feet of the soil surface. Water will rise above the free water table by capillary (adhesion and cohesion forces) action and this zone is called the capillary fringe. Fine-textured soils or organic soils have the greatest capillary rise, and this can be used to subirrigate turf in some situations. Sod farms on flat areas where the water table can be controlled could use subirrigation. Care must be exercised to avoid salt accumulation at the soil surface. Also, turfgrass roots require aeration that could be limited if the capillary fringe produces a saturated root zone. Because root growth may change over a season, the water table may need adjustment.

Subirrigation of golf greens has been attempted from a free water table formed by a drainage barrier. This can present the same problems of salt accumulation, aeration, and cyclic root growth. Other problems that may occur are insufficient capillary rise to meet plant needs and breakage of capillaries during high evaporative demand. These situations are most severe on coarse sands.

A "perched" water table is one that forms due to disruption of internal drainage. Capillary rise of moisture from a perched water table can provide some water needs of turf plants. The perched nature refers to the fact that it is above the normal water table level. The USGA Green Section golf green mix (1) has a perched water table at the pea gravel-coarse sand interface, resulting in higher water-holding capacity for the root zone mix. Once the system starts to drain, however, water movement is rapid. The perched water table is formed by a distinct textural change of coarse sand to pea gravel with a change six to nine times particle-size diameter. As unsaturated water flow occurs through the coarse sand, it does not enter the pea gravel because of poor water-film contact with the pea gravel. A small saturated zone occurs above this interface and drainage does not occur until the saturated zone obtains sufficient depth to break surface tension and to initiate saturated flow.

Sometimes subirrigation of turfgrasses has been attempted using point or line sources of moisture instead of a free water table. Porous tubing buried on a uniform grid under the turf slowly emits water when pressure is applied. This type of subirrigation can exhibit all the potential problems of subirrigation from a uniform water table as well as present some unique problems. Capillary water movement from a free water table is unidirectional, i.e., upward. Capillary movement from a line source is a) uniform in all directions away from the center of the line as long as unsaturated flow occurs or b) if sufficient moisture is present for saturated flow, gravitational forces result in more moisture flowing downward. Very careful spacing of the buried tubing and careful adjustment of the water flow are required to apply adequate water. Too wide a spacing or low-flow rates can result in a striping appearance; too-close spacing or excess-flow rates can cause saturated zones and water leaching. A confounding factor that influences design is that turfgrass-rooting patterns and growth rates are not static but change over the growing season.

Researchers conducting irrigation investigations should take care to avoid reducing the volume of soil for water uptake. Sometimes studies are conducted comparing irrigation regimes at some percentage (i.e., 100, 80, 60 percent, etc.) of field capacity or some percentage of evaporation. An example of irrigating based on percentage field capacity: If water were applied every 4 days at 100, 80, and 60 percent of field capacity, the top 100, 80, and 60 percent, respectively, of the root zone would be brought up to field capacity. This would result in reducing the plant's rooting zone over a period of time and would be contrary to good irrigation practices. A better alternative: Bring all treatments to full-field capacity and allow dry down to 80 or 60 percent of field capacity and then repeat the cycle. This allows wetting of the full root zone and maintenance of a deep root zone. The seasonal ET rates and turf-growth responses will not be the same for the two different methods of irrigation.

When irrigating by percentages of evaporation from a weather pan, one could irrigate every 3 or 4 days, based on 100, 80, or 60 percent of evaporation. This would result in different depths of irrigation. A better method would be to accumulate evaporation to a predetermined moisture level, based on the soil's water-holding capacity, and then irrigating. For example: Assume soil water-holding capacity is 2.0 inches of plant-available moisture in the rooting zone. Because turf responds best by not allowing full depletion, you would wish to irrigate, based on 1.0 inch of soil-moisture depletion. Six days of 0.33-inch evaporation per day would require irrigation on day 3 for the 100 percent evaporation treatment but not until day 5 for the 60 percent of pan evaporation treatment. When irrigation was applied it would be at the 1.0-inch rate for both treatments to allow for the full root zone to be used.

Water availability to plants. In the section on soil water, the energy concept of water was presented. Clearly, the total water content is not a good measure of plant-water availability. Since the plant must compete for soil moisture with the forces of adhesion and cohesion to the soil matrix and solutes, what determines water availability is the water potential or energy status.

The relationship between soil-water potential and water content (volumetric) is shown by a water characteristic curve for a drying soil (fig. 9). Shapes of the curve are similar, but the quantity of moisture retained at a particular potential varies greatly with texture. Water retained at soil matrix potentials (some may be familiar with the older terminology of soil matric suction or tension, which equals matrix potential but with the opposite sign) of 0 to −1.0 bar depends primarily upon cohesion and pore-size distribution. Thus, structure influences moisture retention at high matrix potentials. Below −1.0 bar reten-

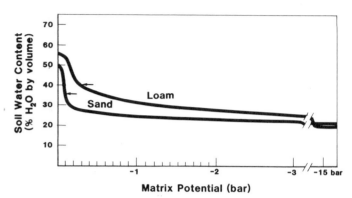

FIGURE 9. Water characteristic curves for a sand and a loam soil. The inflection point (←) on each curve indicates the distribution between capillary and noncapillary pore space for the situation where there is a deep profile without any barrier to drainage.

tion is primarily by adhesion to soil particles and is affected more by texture than by structure.

In a sand soil most of the pores are large and drain at 0 and −1.0 bar; beyond this point the small surface areas of sands result in little moisture retention. Thus, the water characteristic curve of a sand exhibits rapid drainage at high-matrix potentials followed by very little moisture released at the lower potentials. Fine-textured soils have a much wider particle-size distribution. Their curves reveal some rapid drainage at high matrix potentials, due to drainage in the larger pores, followed by a more gradual release of moisture retained in a wide range of similar pores. These different curves would explain why turf on sandy soils exhibits rapid wilting but on a finer-textured soil wilt appears less quickly and does not appear to lead into desiccation as easily.

The water characteristic curve on the moist end of the scale (fig. 10) is very useful in designing a completely modified root zone mix, such as the USGA Green Section method of green construction (1), because it can be used to determine the proper root-zone mix depth. Since the water films around particles are connected with each other, a deeper soil profile will exert greater matrix suction or tension on the soil moisture at the surface.

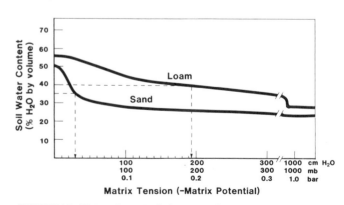

FIGURE 10. Water characteristic curves for a sand and a loam soil. Matrix potential = matrix tension but matrix tension is positive in sign.

Thus, a 25-cm-deep root-zone mix would have a matrix tension of 25 cm H_2O (i.e., the equivalent tension or pull of a water column 25 cm long) on the moisture at the soil surface, while a 50-cm-deep root zone of the same soil would drain to 50 cm matrix tension. This illustrates that the surface-moisture aeration status depends upon the root-zone depth.

The USGA recommends a minimum of 15 percent aeration porosity after a root-zone mix has drained. Using the example of the sand in figure 10, this sand would need to drain to a matrix tension of 31 cm H_2O to achieve a minimum of 15 percent aeration porosity or 15 percent air-filled pore space. This would then require a root-zone depth of 31 cm or 12.2 inches. If the loam were used, it would require a depth of 195 cm (77 inches) to achieve 15 percent aeration porosity. From this illustration, the reader may appreciate why problems arise when athletic field designers use the USGA method for athletic field construction, but substitute a "good loam field soil" for the sand. (Note: This procedure works well for golf greens, football fields, golf tees, and soccer fields.) In the above illustration, a 30-cm layer of the loam soil would have only 3 percent aeration porosity after draining. This would be a very soggy field and a poor media for root growth.

Now that we have reviewed the relationship between soil-water content and moisture potential, the turfgrass grower may wonder how to influence the water potential to make it more available for plants. This should result in more moisture in the range of -0.1 to -2.0 bar, where most of the readily available moisture is obtained. In the previous section we discussed how soil water-holding capacity could be enhanced.

There is one situation where soil-water potential can be directly influenced by a turfgrass manager—the case of salt-affected soils. Normally, the osmotic potential, ψ_o, is minor in humid area soils. In semiarid or arid regions, where soil salt levels may be high, the ψ_o cannot be ignored. One influence of low ψ_o (more negative) is to decrease soil-water potential. Thus, plants take up less water since the driving force for absorption is reduced. Kramer (9) reviewed the results of several investigators and reported that plant roots become less permeable to water when exposed to high salt levels, which further restrict uptake.

If sodium becomes the predominant ion species on salt-affected soils, poor physical conditions can result from soil dispersion. Such soils are termed sodic and have very poor aeration, drainage, and gas exchange.

Removal of the excessive salts will produce a more favorable water potential for plant-water uptake. Leaching salts from the root zone with excess moisture can be effective. This is best done by applying irrigation water in amounts greater than evapotranspiration needs with each irrigation. If leaching is to be successful over a long time, good drainage and a low water table are needed. Water tables within 4 to 6 feet of the soil surface will result in upward flow of water and solutes and salt infusion.

Plant-growth responses cannot be accurately related to soil-water content, but they can be related to soil-water potential, regardless of soil type. All too frequently, water-stress studies are conducted based on arbitrary soil-moisture levels such as: 3, 4, and 6 days between irrigations; irrigating at wilt; irrigating at 5, 10, and 15 percent moisture content; and irrigating at 100, 85, and 70 percent field capacity. It is difficult to relate these soil-moisture levels to plant-moisture stress, except on the particular soil of that study. Soil-water energy status (i.e., water potential) is the best means to characterize soil moisture. If researchers provide the water-potential data, conclusions beyond their immediate soil could be formulated.

In previous sections, theories of soil-moisture retention and movement were discussed as well as practical management techniques to improve moisture-use efficiency. The ultimate goal in water management is to develop an efficient, effective irrigation program. The purpose of the next section is to acquaint the reader with tools or means of estimating turfgrass irrigation needs.

Irrigation programming techniques

As previously observed, the soil-plant-atmospheric continuum (SPAC) is a complex system. Yet the irrigator desires a rapid, easy, inexpensive means of determining impending plant-water stress. Should the irrigator turn to the plant, soil moisture, or some atmospheric parameter to measure and decide when to irrigate? Let us look at each approach by reviewing methods currently found in the literature. Some methods will be useful only to researchers; others will be useful to turf irrigators. In a few instances, the method discussed may not be applicable to turfgrass situations, but it has been presented because of its use on some irrigated crops or in moisture-related research studies.

Monitoring soil-moisture status. At first glance it would appear that monitoring soil moisture would be a reasonable means of determining when to irrigate. When the soil-water potential is high (i.e., soil-water content is near field capacity), soil moisture is not the limiting factor for plant growth. There may be periods, such as a hot, dry afternoon, when the plant exhibits moisture stress under these conditions, but this is due to high evaporative demand, not limited soil moisture. However, as the plant extracts moisture, the soil-water potential decreases and soil moisture becomes limiting. For high quality, rapidly growing turfgrasses this may occur at -0.40 to -0.70 bars. Thus, monitoring soil moisture should aid in irrigation scheduling, but it will not be a sensitive measure of temporary high evaporative demand periods. Methods based on monitoring soil-moisture depletion are summarized below.

Moist-versus-dry soils exhibit different appearances and feel. With some experience, an irrigator can approximate soil-moisture content based on these criteria. This requires an auger to look at several depths. While useful, it is often cumbersome and only accurate within the irrigator's experience.

Use of a soil probe, such as a screwdriver driven into the soil, is similar to the above method. In this case, the grower relates mechanical resistance to moisture content. This method is quick but not very precise.

Tensiometers directly measure soil-matrix potential, which is essentially the same as soil-water potential, except on salt-affected soils. On such soils, tensiometers will not be an accurate measure of soil-water potential. Tensiometers have been used on turf for several years. They are very accurate between 0 to -0.80 bar, but they will not perform below -0.80 bar. Generally two are used with one at 2 to 4 inches and another at 6 to 12 inches, with the depths chosen based on turfgrass-rooting patterns. Irrigation is initiated whenever either tensiometer registers a critical value, for example, −0.45 bar. Irrigation is terminated when the shallowest tensiometer reads 0 (saturation). While tensiometers are accurate, they reflect only conditions at a particular point. Flat, uniform sites are especially well adapted for use of tensiometers. In cold climates they must be removed for winter to prevent freeze damage.

Heat dissipation sensors are commercially available to monitor soil matrix potential. A ceramic tip contains thermometers, with one thermometer surrounded by a heater. This thermometer is heated by a small heat pulse. The rate of temperature change is proportional to water content. These sensors are accurate over a wide temperature range but they are expensive.

Electrical-resistance blocks monitor flow of electricity between wires imbedded inside gypsum or nylon. Blocks are calibrated to relate electrical resistance to soil-matrix potential. Salts interfere, but the gypsum blocks buffer to some extent against the salt influence. Unfortunately for high quality turfgrass situations, moisture blocks are not sufficiently sensitive in the moist range. They are accurate below −2.0 bar and monitor soil-moisture status at a point. Unlike tensiometers, moisture blocks can be left in the soil year-round.

Electrical-resistance probes are generally based on measuring electrical conductivity between two metal probes in the soil. These are very sensitive to salts and are difficult to calibrate versus soil-moisture content. In moisture blocks, soil moisture is transferred via a porous material to the electrodes. This dampens the salt effect as well as standardizes the electrode geometry. With probes neither factor is easily controlled without care by the operator. As a result, electrical probes are difficult to calibrate versus soil-water content.

Neutron probes, gamma-ray attenuation, and soil psychrometers are available to measure accurately soil-water content (first two) or soil-water potential (latter). However, these are very expensive and, in the case of psychrometers, difficult to calibrate since each psychrometer requires calibration for soil-water potential and temperature gradients. Psychrometers are not as accurate in sandy soils. Also, neutron probes and gamma-ray attenuation techniques do not lend themselves to measuring soil moisture in the surface few inches, which is important for turf situations. These methods would be primarily of interest to research scientists.

Recent developments in time-domain reflectometry (TDR) may provide a practical tool for irrigation (16, 17). It appears that the dielectric constant of soil (measure of the extent of polarization in an electrical field) is primarily related to its water content. This technique measures the soil-dielectric constant, and this can be done in discrete zones in the soil, including the soil surface layer. A commercial unit has recently become available. Such a unit may be programmed similarly to a tensiometer. Certainly this could be useful in research studies for monitoring water flow. It appears to be sensitive over a wide soil-moisture range, but units are expensive.

Monitoring atmospheric demand. Using meteorological data to determine plant-water requirements is another alternative for irrigation schedulers. This is based on the fact that atmospheric conditions dictate to a large degree the rate of soil-water loss by evaporation and transpiration. However, the plant can reduce depletion by stomatal closure and when soil moisture declines, further extraction is at much lower rates than potential evaporative demand would indicate.

Both mathematical and weather pan approaches have been used by scientists to estimate ET of plants as related to evaporative demand. Before discussing these two methods, let us review the basic, common theory behind both approaches. The first step is to estimate potential ET (ETp), where ETp = maximum evaporation and transpiration from a short, green reference crop, assuming soil moisture is not limiting. Once a potential ET (ETp) is obtained, it is related to actual crop ET (ETc) by a crop coefficient factor (k), which depends on crop age, growing season, crop characteristics, soil-moisture status, and prevailing weather conditions.

$$ETc = k\ ETp$$

Once a grower knows ETc on a daily basis, the grower can monitor turf-water losses and use this to schedule irrigation. Ideally, the grower would collect this data (i.e., the ETp and k factor) for a specific site. Another possibility would be to collect the ETp data at several sites within a region and then announce the ETp each day. Growers would then use this as a guideline for their irrigation. In the latter method, the region would be zoned into similar areas and growers provided with estimated crop coefficient factors for their zones. California is reportedly initiating such an approach to provide irrigation guidelines for homeowners as well as professional turf managers (Victor A. Gibeault, personal communication).

As noted previously, ETp can be estimated by either mathematical procedures, or weather-pan data. Doorenbos and Pruitt (6) and Taylor and Ashcraft (15) provide good discussions of the different mathematical procedures for estimating ETp from climatic data. The methods can be classified as: a) energy balance methods, b) mass transfer, or aerodynamic methods, c) combination methods—where aerodynamic and energy balance concepts are combined, and d) empirical equations with various levels of theoretical basis. The empirical procedures (Penman, Thorthwaite, Blaney-Criddle, Jensen–Haise, locally developed equations, etc.) have been the most useful (6, 15). These equations utilize various climatic data—air temperature, solar radiation, relative humidity, wind factors—to calculate ETp daily, weekly, or monthly.

Several problems have been encountered when using empirical procedures for estimating ETp. No universal equation has evolved, but some seem to be more useful in certain climatic areas, such as arid, semiarid, cool-humid, etc. For turfgrasses, daily ETp data are necessary, but all of the equations are most accurate over longer periods. Many times climatic data are not available on a routine basis to the grower and monitoring equipment is expensive. Also, calculations can be complex, but pocket calculator programs can be developed.

The weather-pan procedure for estimating ETp has potential use for both researcher and grower. Evaporation of free water from a weather pan integrates the effects of radiation, temperature, wind, and humidity. These are the same climatic factors that influence crop-water use (6). This technique has been utilized successfully on turfgrass (18).

Evaporation from a weather pan (Epan) can be correlated to ETp by a pan coefficient (Kp).

$$ETp = Kp\ Epan$$

Sometimes individuals have used a constant pan coefficient for all climatic conditions. This is not a good practice since Kp changes with different humidity, wind conditions, and type of weather pan. Tables for Kp under different wind and humidity conditions and for two different weather pans are available (6). The U.S. Weather Bureau pan is most commonly used.

Once the daily ETp value is obtained, it is then related to actual crop evapotranspiration by the formula:

$$ETp = k\ ETc$$

In some cases researchers have converted directly from Epan to ETc. This is probably not the best situation, since it combines the coefficients Kp and k.

$$ETp = Kp\ Epan = k\ ETc$$

$$ETc = Kp/k \cdot Epan$$

As we have just seen, Kp is not a single unique value, but it is influenced by weather conditions.

Regardless of whether an ETp value is obtained from weather pan or mathematical procedures, it must be related to actual turfgrass evapotranspiration, ETc, by the crop coefficient, k. For crop plants in general, crop coefficient (k) is influenced by type of crop, stage of growth, climatic conditions, and soil moisture. For a mature turfgrass sod, major factors affecting k would be type of grass, time of year (turf-growth cycle), climatic conditions, and soil-moisture status. The soil-moisture aspect can often be ignored on well-irrigated turf. If irrigation is not initiated until the turf undergoes at least moderate moisture stress ($\psi_s \approx -2.00$ to -4.00 bar), then the k value is different from less limiting soil-moisture conditions. While k values can be taken from tables (6) or derived from experimental results, the most accurate are obtained experimentally at the specific location, i.e., climatic zone. This can be done by using actual ETc values observed in lysimeter studies and relating these to ETp from weather pan or mathematical origin. From this discussion, it should be apparent that k values must be seasonally adjusted.

We have reviewed two procedures for monitoring atmospheric demand as a way to provide irrigation guidelines. Another alternative for researchers can be lysimeters to measure directly crop evapotranspiration. These can be simple in design, such as buried lysimeters made from PVC pipe and removed for weighing, or they can be elaborate with built-in weighting devices. To reflect growing conditions accurately, lysimeters should be placed carefully in sodded areas. In studies where root-system distribution would influence the ET rates of a grass, such as comparing two tall fescues under limiting soil moisture, lysimeters may not accurately reflect a true field situation since full-root development may be restricted. This would be a problem on lysimeters that are shallower than the normal rooting depth. Another potential problem is impeded drainage, which could result in a positive pressure potential. Such ponded water can influence ET rates. Researchers using lysimeters should carefully consider their irrigation regimes, as discussed in the section on water-holding capacity, if the water-use data is to reflect accurately practical field situations.

Johns, Van Bavel, and Beard (8) have reported on the determination of resistance to sensible heat flux density as a procedure to determine turfgrass ET. This method allows rapid determination of ET in the field, using an infrared thermometer and a net radiometer. While this procedure should prove useful in ET rates at a point in time, such as in a study comparing tall fescue cultivars at a point in time, it does not provide a daily ET value. For irrigation scheduling purposes, a daily water-use rate is necessary.

Monitoring plant-water status. Ideally, the grower would like to measure some parameter of the plant that would indicate impending moisture stress. Such a parameter would integrate all the soil-atmospheric-plant factors. Researchers have investigated leaf-water potential and stomatal diffusive resistance, but these techniques are not refined to the point of usefulness as irrigation criteria. Also, they require many point measurements to reflect accurately turfgrass canopy conditions.

A simple plant parameter that aids irrigation programming is wilt. One would like to irrigate before wilt symptoms appear, since irrigating at wilt does result in some turf deterioration. However, if a grower uses wilting of indicator spots, this will indicate the relative moisture status of adjacent areas. On most turf sites, some area consistently exhibits wilt first. This observation will require experience, but it can be useful.

Infrared thermometry allows nondestructive monitoring of turfgrass canopy temperatures. It also averages together many plants. While under no moisture stress, turf canopy temperatures are near (some species are slightly below and others slightly above) ambient air temperatures. As stress is imposed, turf canopy temperatures rise above ambient air temperatures. Monitoring and relating these temperature changes to plant-water requirements could aid in programming irrigation. Researchers have presented several articles relating to this technique and its potential (13). A project is currently being conducted by the author on the use of infrared thermometry for turf irrigation scheduling.

Combination method. Previously we looked at the soil-plant-atmospheric system as a budget system with inputs, outputs, and a certain reserve of moisture at any one time (fig. 7). Irrigators should keep this concept in mind when fine-tuning irrigation programs. By measuring or estimating each component, the irrigator can more carefully use any of the techniques (soil, atmospheric, plant-water status) just discussed.

Also, irrigation personnel—whether turf managers or researchers—need to answer the question, "How much water should I apply?," not just the question "When should I irrigate?" The techniques may guide in answering the latter question but not necessarily the former. How much moisture to apply requires knowledge of the depth of rooting, degree of moisture extraction, and soil-water holding and drainage relationships. Adjusting irrigation application rates to the existing soil and plant-rooting conditions will substantially contribute to water conservation.

Literature cited

1. BEARD, J. G.
   1982. Turf management for golf courses. Burgess Pub. Co., Minneapolis, Minn. 642 pp.
2. ———
   1973. Turfgrass: science and culture. Prentice-Hall, Inc., Englewood Cliffs, N.J. 658 pp.
3. BOURGOIN, B, and P. MANSAT
   1979. Persistence of turfgrass species and cultivars. J. Sports Turf Res. Inst. 55:121–140.
4. CANAWAY, P. M.
   1978. Trials of turfgrass wear tolerance and associated factors—a summary of progress 1975–1977. J. Sports Turf Res. Inst. 54:7–14.
5. CARROW, R. N.
   1980. Influence of soil compaction on three turfgrass species. Agron. J. 72:1038–1042.
6. DOORENBOS, J., and W. O. PRUITT
   1977. Crop water requirements. FAO Irr. and Drain Paper 24. Food and Agric. Organ. of the United Nations, Rome. 144 pp.
7. HILLEL, D.
   1980. Fundamentals of soil physics. Academic Press, N.Y. 413 pp.
8. JOHNS, D., C. H. M. VAN BAVEL, and J. B. BEARD
   1981. Determination of the resistance of sensible heat flux density from turfgrass for estimation of its evapotranspiration rate. Agric. Meteorol. 25:15–25.
9. KRAMER, P. J.
   1969. Plant and soil water relationships: a modern synthesis. McGraw-Hill Co., New York, N.Y. 482 pp.
10. MADISON, J.
    1971. Principles of turfgrass culture. Van Nostrand Reinhold Co., New York, N.Y. 420 pp.
11. O'NEIL, K. J., and R. N. CARROW
    1982. Kentucky bluegrass growth and water use under different soil compaction and irrigation regimes. Agron. J. 74:933–936.
12. O'NEIL, K. J., and R. N. CARROW
    1983. Perennial ryegrass growth, water use and soil aeration status under soil compaction. Agron. J. 75:177–180.
13. PHENE, C. J., and E. C. STEGMAN
    1981. Irrigation scheduling for water and energy conservation in the 80's. Proc. Am. Soc. Agric. Eng. Irr. Sched. Conf. Pub. Am. Soc. Agric. Eng. St. Joseph, Mo. ASAE Pub. 23–81. 231 pp.
14. PHILLIP, J. R.
    1966. Plant water relations: Some physical aspects. Ann. Rev. Plant. Physiol. 17:245–268.
15. TAYLOR, S. A., and G. L. ASHCRAFT
    1972. Physical edaphology—the physics of irrigated and nonirrigated soils. W. H. Freeman and Co., San Francisco, Calif. 533 pp.
16. TOPP, G. C., J. L. DAVIS, and A. P. ANNAN
    1982. Electromagnetic determination of soil water content using TDR: i. applications to wetting fronts and steep gradients. Soil Sci. Soc. Am. J. 46:672–678.
17. ———
    1982. Electromagnetic determination of soil water content using TDR: evaluation of installation and configuration of parallel transmission lines. Soil Sci. Soc. Am. J. 46:678–684.
18. WADDINGTON, D. V.
    1969. Soil and soil related problems. *In* Turfgrass Science. A. A. Hanson and F. V. Juska (eds.). Amer. Soc. of Agron. Monograph 14. pp. 80–129.
19. YOUNGNER, V. B., A. W. MARSH, R. A. STROHMAN, V. A. GIBEAULT, and S. SPAULDING
    1980. Water use and turf quality of warm-season and cool-season turfgrass. *In* Proc. 4th Inter. Turf. Res. Conf. Ontario Agric. Coll. and Inter. Turf. Soc. pp. 251–257.

Practicum

9 Irrigation systems for water conservation

JEWELL L. MEYER

Jewell L. Meyer received a B.S. degree from the University of California at Berkeley and an M.S. degree from UC Davis, and he is a registered professional engineer in California. He became a farm advisor and irrigation specialist in 1954 with UC in the San Joaquin Valley, and since 1978 he has been located at UC Riverside.

BRUCE C. CAMENGA

Bruce C. Camenga has had 20 years' experience in the irrigation industry, including responsibility for hydraulic design education for a major turfgrass irrigation systems manufacturer. Presently, he is owner/operator for Oasis Irrigation Supplies in Riverside, California.

Water conservation must be a planned objective in the original design of a turfgrass irrigation system. Although all irrigation systems are a series of compromises during the design process, long-term water savings can justify the careful selection of new, high-technology controllers, valves, sprinklers, and safety devices.

Budget considerations should be judged by the amount of water saved over the life of a project. Dollar values can be assigned to water savings with the use of the following new equipment:

- Low-precipitation-rate sprinklers,
- Matched-precipitation sprinklers,
- Soil-moisture sensors,
- Rain sensors,
- Wind sensors,
- Drainage check-valves, and
- Repeat-cycle and variable-watering programmed controllers.

Estimates of water savings through runoff control and high uniformity from good sprinkler performance are measurable by the percentage of uniformity. In design, this uniformity is calculated from manufacturers' specifications. "Can" tests are made in the field to evaluate actual precipitation at equal spacings in the water pattern. The calculated uniformity is called CU and is given as a percentage. Watering times are usually increased by the percentage of uniformity of the sprinkler system. Actual watering time is based upon the decimal percent less than 100 by dividing theoretical watering time or ET by the CU. A sprinkler pattern or system operated at 60 percent uniformity must deliver one-third more water than a system with 80 percent uniformity to achieve the same green, nonstressed turfgrass. Any excess water is lost to runoff or deep percolation and represents a dollar loss.

The application rate of sprinklers must be as low as practically obtainable. Ideally, the application rate should match the soil intake rate. Repeat-cycle controllers are now available for systems on low-intake-rate soils. Irrigation cycles are designed to shut down when runoff is about to occur; a repeat cycle of short duration is used after water has penetrated the soil. When the appropriate depth of wetting occurs, sensors can interrupt the preprogrammed water cycle.

Revolving or rotating sprinklers significantly reduce the rate of water application. The single stream or multiple streams pass over the turf at a slow rate. Fixed-orifice spray heads all have much higher application rates than rotating heads and in general are much less uniform. Matched-precipitation-rate sprinklers that allow the manager to operate one-fourth-, one-half- and full-circle sprinklers on the same line drastically reduce the cost of valves and pipelines, and they usually improve overall efficiency. Only matched-precipitation sprinkler heads and rotational heads should be considered for water conservation designs.

Management of a water-conserving system must follow the principles of the original or as-built design. Sprinklers must continually be examined for application rate, damage, and vertical position. Turf "can" tests should be performed to determine application rate before brown spots occur.

Controllers should be programmed according to water use. Programming watering times less frequently than monthly will result in wasted water in the fall and winter and plant stress in the spring and summer. Ideally, weekly programming will save the most water. In addition, tensiometers have to be serviced each week.

9 Irrigation systems for water conservation

Jewell L. Meyer and Bruce C. Camenga

The degree of interest in landscape water conservation is directly proportional to the cost and/or availability of water. The less water available and the more costly it is, the more concern is evident for water conservation.

This is not to say water conservation should not have been enthusiastically endorsed during preceding decades, but drought conditions, costs of water treatment, and escalating energy costs have caused water conservation in the landscape to become especially significant in recent times.

Energy costs greatly affect water because it takes energy to move water from where nature has deposited it to where it is needed—if only from a well in the backyard to the house, or from mountain-fed rivers to dry plains several hundred miles away.

When water becomes scarce, there are very real challenges for the landscape industries in conserving water supplies by making optimum use of available water. To prevent the rapid political remedy of eliminating irrigation of turf and landscapes, the challenge is to perfect more efficient methods and equipment to irrigate turfgrass and plant materials. Discussions in this chapter will cover:

- Budget considerations,
- Efficiencies of coverage (R_a),
- Efficiencies of distribution (D_u),
- Site considerations,
- Design considerations, and
- Management of the irrigation system.

This chapter deals with water conservation of sprinkler systems. Drip-and-trickle irrigation is a special method for conserving water in trees and shrubs, but, at this time it does not have universal use in primarily turfgrass areas and is not discussed here.

Budget considerations

Every irrigation system is a study of compromises. The best compromises are those that do not adversely affect the quality of the irrigation system but still keep costs of the system within a budget. This is a deceptive statement because, usually, when budgetary considerations are made, a greater emphasis is placed upon up-front or immediate costs. Actual operating costs receive little or no consideration. Operational costs cannot be overlooked, because annual costs of irrigation system management are becoming more and more a financial burden to private and public developments.

Design compromises are possible that do not adversely affect the quality of the system. These compromises deal with the level of sophistication of the system and, often, with convenience. When an irrigation system is planned, water conservation should be one of the many goals. The irrigation system should be a tool for the turfgrass manager; however, sometimes the manager wants special features that will boost his ego, but will prove later to have limited practical application. Cost justification should be used for every feature with a price tag! Compromises can be made to aid the budget without harming the true capabilities of the system. Then, possibly, funds can be used to finance features that will justify costs based on the amount of water and money saved over the life of the project.

The new irrigation system features which are now available and have been shown to pay for their initial costs in as few as two years are as follows:

- Moisture-sensing devices,
- Rain-sensing devices,
- Wind sensors,
- Low-precipitation sprinklers,
- Matched-precipitation sprinklers,
- Excess-flow sensing devices,
- Devices to prevent low-head drainage,
- Master valves,
- Special controller circuits to provide varying watering schedules,
- Repeat cycling controllers, and
- Solid-state time accurate controllers for accuracy in timing.

These system features are costly, but if they are viewed in light of how much they will save in operational costs, their acceptance will increase tremendously.

When water costs on a golf course are over $1,000 per month, it is easy to justify spending another $1,000 to $3,000 on equipment that can reduce the amount of water used by 5 or 10 percent.

Efficiency of coverage

When considering water conservation, the two most important components of an irrigation system are the sprinklers and the controller. The sprinklers determine the uniformity of coverage, but they are too often selected on the whim or plan of the irrigation system designer. Thus, although not a part of the irrigation system, the designer becomes the most important ingredient in planning the irrigation system. The relative problems of sprinkler types is discussed in the next subject, efficiencies of distribution; the role of sprinkler design, usage, and coverage is addressed there.

Control of overthrow onto areas that do not require water, control of application rates, and control of frequency are important considerations in irrigation design. Inefficient systems have sprinklers which overthrow onto sidewalks, driveways, or other paved areas, or in turfgrass areas they throw water into shrubs and ground covers. This is not because part-circle sprinklers are not available; it is because fewer sprinklers are often used, translating to less material cost for the system, but more wasted water and higher operational costs.

The rate of water application by sprinklers is very important in water conservation. Sprinklers with high rates of application are prone to create runoff or percolation beyond the root zone of the turfgrass. Some conservationists would dispute the fact that this causes wasted water, as water from runoff or deep percolation is retained in the general area. Runoff water and water that percolates below the root zone have been paid for by the turfgrass manager and are not beneficial to the turf; therefore, they must be considered lost to his project. The in-pattern rate of application (R_a) of an irrigation system can be calculated by using the following formulas:

$$R_a = \frac{\text{sprinkler output}}{\text{sprinkler spacing}}$$

$$R_a = \frac{\text{gpm} \times k}{\text{head spacing}}$$

$$R_a = \frac{\text{*gpm of full circle head} \times 96.3}{\text{head spacing ``a''} \times \text{head spacing ``b''}}$$

in an equilateral pattern

$$R_a = \frac{\text{*gpm of full circle head} \times 96.3}{\text{head spacing ``a''} \times \text{row spacing ``b''}}$$

or

$$R_a = \frac{\text{gpm* of full circle head} \times 96.3}{\text{head spacing ``a'' squared} \times .866}$$

*To determine application rate of a half-circle sprinkler in pattern, multiply gpm by 2; for a quarter-circle sprinkler, multiply gpm by 4.

(R_a) is very important. Almost all available sprinkler heads have a higher rate of application than the percolation rate of many soils.

Fixed, single-orifice sprinkler heads, commonly called spray heads, usually have a rate of application exceeding 1½ inches per hour. As the trend to use spray heads with a little longer radius continues, application rates become exponentially higher. More water must be used to gain greater radius of coverage. The application rate can then become intolerable, exceeding 2 inches per hour.

The following examples show the rate of application of a common spray head. An increase of 1 foot in radius increases the application rate 25 percent and only increases spacing by about 10 percent.

Half-head, 13-ft radius, 15-ft triangular spacing, 2 gpm

$$R_a = \frac{2 \times 2 \times 96.3}{15 \times 15 \times .866} = 1.44 \text{ in per hr}$$

Half-head, 14-ft radius, 17-ft triangular spacing, 2.5 gpm

$$R_a = \frac{2 \times 2.5 \times 96.3}{17 \times 17 \times .866} = 1.9 \text{ in per hr}$$

The use of revolving or rotating sprinklers can significantly reduce the rate of application. The stream or streams of water pass over only a portion of the ground to be covered at any one time. Less water is used and is concentrated in a stream that is thrown farther. The application rate of full-circling rotating sprinklers is usually about 1/2 inch per hour. Using the preceding formula for a typical large turf-type rotating sprinkler that throws water in a 60-foot radius at a rate of 20 gpm, is spaced at 60 percent of the diameter of the head (normally recommended maximum of most manufacturers) in a triangular pattern, the Ra is:

$$\frac{20 \times 96.3}{72 \times 72 \times .866} = .43 \text{ in per hr}$$

This shows rotating sprinklers have a desirable lower rate of application. The more efficient choice of rotational or spray heads would be the rotating sprinkler when the site is suitable. Rotating sprinklers are more expensive, however, from a unit-cost basis or from a total equipment cost. This has precluded selection of rotating sprinklers by those with a limited viewpoint. Operational costs, a factor in the selection of sprinklers, usually offer water savings and justify selecting the rotating-type sprinkler.

Design engineers can calculate the runoff water potential when determining the selection of high application-rate spray heads versus lower-rate rotational sprinklers.

Cost justification should consider the length of time it will take for water-cost savings to meet additional equipment expense. The time period will vary due to the water demands of the turfgrass, climatic conditions, and length of watering season. The longer the season, the higher the evapotranspiration rate; and, the drier the climate, the greater the justification.

The following example shows how to justify the extra cost of a more efficient irrigation system. It is for a 1-acre area with an estimated savings of 5 percent of watering costs due to runoff reduction. If a project larger than 1 acre is under consideration, multiply the cubic feet per acre of water saved by the number of acres in the project.

Example of cost justification:

One acre with 1½ inches per week applied in summer months, 1 inch per week in spring and fall months, and water costs of $300 per acre-foot.

$$(1.5 \text{ in} \times 12 \text{ wks}) + (1 \text{ in} \times 24 \text{ wks})$$

$$(1.5 \times 12) + (1 \times 24) = 42 \text{ acre-in of water per yr}$$

$$42 \text{ acre-in} \times 27{,}154 \text{ gal/acre-in} = 1{,}140{,}468 \text{ gal}$$

$$1{,}140{,}468 \text{ gal} \times .05 \text{ water savings} = 57{,}023 \text{ gal}$$

$$57{,}023 \text{ gal}/7.48 \text{ gal in ft}^3 = 7{,}623 \text{ ft}^3$$

$$\frac{7{,}623 \text{ cu ft}^3}{1{,}000 \text{ ft}^3} \times \$7/1{,}000 \text{ ft}^3 = \$53/\text{A savings per yr}$$

Frequency of water application is another cost factor. The axiom is to water infrequently and deeply for normal crop growth. For turfgrasses, however, deep watering below the root zone is lost water. High application rates and low permeability of many soils usually cause high water runoff.

Again, it is necessary to compromise between the theoretical best method and the most practical method. It is essential to determine how long water can be applied to a specific turfgrass area with a given soil type and slope before runoff occurs. Watering must be discontinued before runoff occurs. This does not mean that enough water has been applied to the turf to make up for water use through evapotranspiration. The practical solution to this challenge can be multicycling the irrigation system. Water is applied for a period not to exceed the point at which runoff begins. Irrigation is discontinued long enough for applied water to penetrate into the soil. Then, another short application is administered. Depending on the soil and turfgrass condition, these cycles can be repeated as many as four or five times in a day. An analogy would be calories needed to sustain a level of physical activity for an individual. If x calories are needed to prevent deterioration, it doesn't necessarily follow that all calories must be obtained at the same time. Several intakes throughout the day are better than one large intake. The end result is accomplished without overeating.

Runoff is also caused by low-soil porosity and the condition of the turfgrass. In heavily matted thatch, water has a more tortuous path in getting to the soil, and runoff occurs more quickly than in dethatched turfgrass. Proper maintenance of turfgrass assists water conservation.

The multicycling concept of irrigation is used in combination with moisture-sensing devices to maintain the optimum level of moisture in the soil. Sensors allow watering at a predetermined starting time when the moisture level has dropped to a preselected point. If the frequency of application is for long time periods, the roller coaster of too little and too much moisture may occur. When predetermined watering starts are frequent and short, water will be applied only to the point of satisfying the moisture sensors and the water needs of the turfgrass.

Multicycling with moisture sensing keeps the soil water level closer to optimum conditions. These conditions are accomplished by using several programmed cycles per night and every-night programming. This does not mean that the irrigation system will run every night nor for every cycle programmed on the controller. The system will only run when the moisture level in the soil is depleted to a predetermined point before stress occurs. Sensors can be placed at two depths in the turfgrass, shallow and deep root. A moisture level decrease in either zone will trigger the irrigation cycle. Both zones must be at a satisfactory moisture level before irrigation starts are cancelled.

Mismatching areas of water need is another problem in irrigation systems design. There is a tendency to sacrifice watering precision, if an extra valve or an extra station on a controller is needed. Equipment cost savings is usually the excuse for this determination. When water conservation is an irrigation system goal, this type of decision making will never accomplish precise watering.

Precise watering to meet plant requirements involves sectioning the irrigation system—that is, deciding which sprinklers and/or valves to run together or separate on a controller station.

No rational designer would allow sprinklers on a golf course green and a rough to be run at the same time, for the same amount of time, and at the same frequency. The turfgrasses in these two areas are different, as are their soil and playing conditions. Some designers, however, do not hesitate to run residential areas, parks, or industrial complexes on the same program even though they differ in soil, turf, usage, slope, drainage, sun, and shade characteristics. In the extreme, turf areas and ground cover or shrubbery areas are watered at the same time, even though each has different water requirements.

The proper design of a landscape irrigation system must match and complement the landscape. The system is for the protection and stimulation of the landscape; without the landscape, the system would be unnecessary.

Efficiencies of distribution

Distribution of water from sprinklers is very important to water conservation. Water distribution in the area being irrigated is also important and can differ from distribution patterns of an individual sprinkler.

Most sprinklers cover a circular area, and because water distributed in the circle decreases from the center point to the outside of the circle, each sprinkler must overlap another—this is referred to as percentage of coverage or overlap. Many manufacturers recommend coverage as a percentage of the diameter of the sprinkler with the type of spacing pattern used. There is little evidence that the radii shown for their equipment are realistic in actual usage. Commonly recommended spacing is 60 to 65 percent of the diameter of the sprinklers for equilateral spacing and 55 to 60 percent of the diameter of the sprinklers for square spacing. Caution: This type of recommendation is for "no-wind" conditions. Some manufacturers offer a derating percentage that increases with the velocity of the wind up to 5 to 10 mph. At 5 to 10 mph wind, spacing is 50 to 55 percent of the diameter of the sprinklers for triangular spacing and 45 to 50 percent of the diameter of the sprinklers for square spacing. The 50 percent figure is 100 percent coverage or overlap.

These recommendations assume distribution of the sprinkler is near-perfect in "no-wind" conditions and remains nearly undistorted, although shortened, when there is wind from any direction. This, of course, is an oversimplification. The recommendation for coverage or overlap should be conditioned on the profile of the sprinkler and should be tied to the optimum operating pressure for that particular sprinkler (fig. 1). Higher or lower than optimum pressure will distort the sprinkler's profile. This distortion causes unequal application of water in the pattern. When water is unequally applied, wasted water is the end product.

Wind direction causes lowered application rates in the up-wind direction. Prevailing winds should be considered in irrigation system design. Spacing should be reduced perpendicular to the wind. Merely knowing that a 3-to-5-mph wind is usually blowing during irrigation times is not enough, and reducing the spacing in all directions may be unnecessary and unwise, as it may deteriorate the efficiencies of distribution even more. On-site inspections and examination of historical wind patterns help in critical windy areas.

SPRINKLER PROFILES

EFFECT OF DIFFERENT PRESSURES ON DISTRIBUTION PROFILE

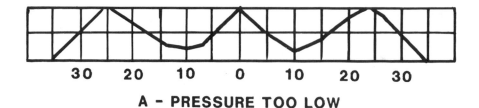

A - PRESSURE TOO LOW

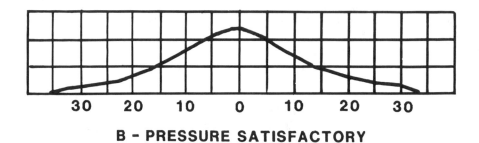

B - PRESSURE SATISFACTORY

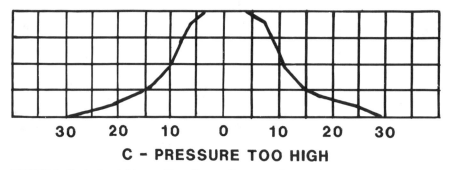

C - PRESSURE TOO HIGH

FIGURE 1. Typical sprinkler profiles of low, optimum, and excessive pressures

Low-precipitation-rate sprinklers have been discussed earlier because they tend to decrease runoff. They increase overall efficiencies of the irrigation system because they generally have a lower differential between the wettest and driest spots in coverage; therefore, they increase distribution uniformity.

The matching of precipitation rates for part- and full-circle sprinklers can also help increase uniformity of an irrigation system.

Quarter-circle sprinklers are often used in corners, half-circle sprinklers on borders, and full-circle sprinklers in open areas. There were only two ways to keep water distribution even. If all the sprinklers have the same gallonage delivery rate, then the quarter-circle sprinklers must be run independently for half the time of the half-circle sprinklers. If the gallonage delivery is the same and the area is one-half that of a half-circle sprinkler, the application rate of the quarter-circle sprinklers is twice that of the half-circle sprinkler. Full-circle sprinklers must be independent of the quarter- and half-circle sprinklers, as their area of coverage is different, although their gallonage delivery rate is the same. Sectioning sprinklers in systems with many borders becomes difficult and costly if quarter-circle sprinklers are placed on separate valves from half-circle sprinklers, and half-circle sprinklers cannot be on the same valves as full-circle sprinklers.

The new way to achieve uniform delivery is to use sprinklers with matched precipitation rates (fig. 2). A full-circle sprinkler's precipitation rate, matched with those of a half- and a quarter-circle sprinkler, will be twice the rate of the half-circle sprinkler and four times that of the quarter-circle. Using matched sprinklers with similar radii allows all three patterns of sprinklers to be run on the same valve and/or controller station.

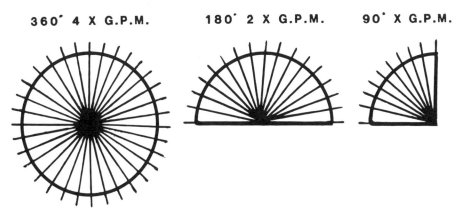

FIGURE 2. The sprinkler patterns of matched precipitation rate heads

Site considerations

Site considerations are sometimes overlooked in irrigation design, especially where system efficiency and water conservation are emphasized.

On-site conditions that can affect the efficiencies of the system include, but are not limited to: slopes, varying soil conditions, types of landscape materials, and varying evapotranspiration rates due to sun conditions and plant variations.

Many of these were covered in the section, "Efficiency of coverage," which discusses matching the irrigation system to the landscape. In some cases, site considerations go beyond matching the system to the landscape. Slopes are mentioned because their runoff point is reached sooner than it is on level terrain. Low-head drainage on slopes also becomes important. When a sprinkler-distribution line (lateral) runs up or down a slope to feed several sprinklers, the line water will drain to the lowest sprinklers each time the control valve is shut off. The valve may be at the low point, but the higher elevation lines will drain back out of the lower-elevation sprinklers. This drainage causes a soggy condition around the lower-elevation sprinklers that can become a quagmire if there is foot or vehicular traffic in the area. The drainage problem can be overcome by artfully selecting the sprinklers to be run on each valve or by including check valves at each sprinkler or automatic drain valves at all lower elevation sprinklers. Check valves will hold the water in the lateral lines, whereas automatic drain valves will discharge the water belowground in gravel-filled sumps (fig. 3).

Design considerations

Many decisions have to be made by the irrigation system designer; there are, however, user and management decisions the designer must consider when formulating a plan to irrigate a landscape efficiently.

The facility's primary use must be considered by the irrigation system designer. Landscape use determines the time available for irrigation and will sway decisions on how and when the irrigation can be effectively and efficiently achieved. Golfers do not appreciate paying greens or membership fees and then having to play through sprinklers or on freshly watered turf. Irrigation of golf courses is scheduled during the nonuse time—generally from 7 p.m. to 7 a.m.; thus, only 12 hours are open for irrigation. The highest plant water use, ET, occurs when the playing hours are the longest; an insidious trick nature played on groundskeepers gives players more daylight-use hours and a higher ET rate at the same time.

Parks, schools, cemeteries, memorial parks, industrial and commercial complexes, and even residential developments also have times when irrigation cannot be permitted. Irrigation must occur when the facility is not being used. Residential areas are generally not such a problem as there is plenty of time to irrigate. Schools and industrial and commercial complexes are also not too demanding as long as certain high-use hours are avoided. Parks often pose a problem because their hours of use are increasing. Baseball, softball, football, soccer, jogging, and other uses extend from early morning to as late as midnight in many public parks. Shortening available time for watering can increase the irrigation system cost because the total water delivery rate must be increased to accomplish proper irrigation. Days and hours per day available for watering and the amount of water to be applied form an "irrigation interaction relationship triangle." When any one portion of the triangle is affected, another portion is also affected. Generally, the fixed portion of the triangle is the amount of precipitation required by the turf. If watering time is decreased (by reducing hours of watering or days available for watering), the water supply must increase. Conversely, if the water supply is to be decreased, watering time must increase. In either case, if an increase in supply or time is not affected, the amount of water delivered to the turf is reduced and the turf will quickly reach or exceed the stress point, causing deterioration in turf quality.

Automatic programmable systems have been almost exclusively discussed in this chapter; however, manual systems and

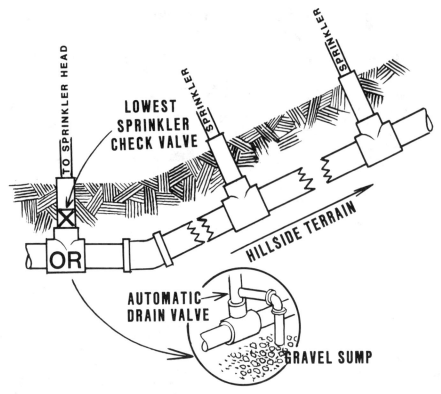

FIGURE 3. A diagram showing placement of check valves or drainage valves with a small sump

quick coupling systems are also used with some degree of success. The same methods to conserve water can, theoretically, be used on manually operated or quick-coupling systems, but other manual types have become impractical.

A quick-coupling system's irrigation lines are permanently buried, and its valves are installed in each desired sprinkler location. Quick-coupler keys with sprinklers are inserted into the number of valves the system can support for irrigation. The sprinklers are run for a predetermined time and then are moved to another location. An irrigator has to move the sprinklers manually, and multicycling is virtually eliminated as a method of conserving water. Automated moisture sensing is also eliminated as a conservation method, although manual moisture-sensing techniques can still be employed to determine when to water and how much. Accuracy is sacrificed with a quick-coupling system. Although a human can be as accurate as a timing device, it rarely happens consistently in the field.

Material cost of a quick-coupling system is less than that of an automatic system, but labor costs often offset savings. High labor costs continue beyond the point of equaling the material and continue the longer the system is used.

The basic differences between a manual system and an automatic system are the lack of automatic control, the need for manual labor, and limitations of accurate controls.

Automatic systems vary in the degree of automation. Some purists claim that many systems labeled automatic really are not automatic, but simply work on a preprogrammed cycle. The term "automatic" should refer to the method by which the valves are operated, i.e., the valves automatically open on a signal from a control unit. This is similar to the automatic transmission of an automobile in which the transmission will automatically shift from gear to gear while running in a forward position based on a preselected range of function. The transmission does not sense when to move forward or backward any more than a conventional automatic irrigation system can sense when to water and for how long. Features can upgrade an automatic system with moisture sensing as an integral part of the system, but even the proper use of moisture sensing relies on preprogramming of the moisture sensors and the times of day at which the controller is allowed to run.

Equipment selection must be based on those goals to be accomplished with the irrigation system. Features of a system must be justified on their abilities to aid in the efficiencies and uniformity of the system.

Features of irrigation equipment that can assist in achieving water conservation are:

Moisture sensing. Use of sensors in both the shallow and deep root zones of representative areas of the landscape can cancel predetermined start-and-run times of the system when a preselected level of moisture is available in both root zones. The equipment may work on vacuum or electric conductivity.

Excess flow-sensing device. This device can sense when the water flow in a system is abnormally high, due to a broken line or missing sprinkler, and will shut the system down, usually giving a warning so the human element can locate the problem and correct it. With today's sophisticated electronics, the problem can be pinpointed to the station where the excess flow occurs and sometimes to the actual location of the problem.

Low-precipitation sprinklers. Manufacturers are aware of the advantages of low-precipitation sprinklers. Irrigation system designers are beginning to use this type of sprinkler to help in water conservation (and sometimes to save initial and long-range costs). Low-precipitation sprinklers can reduce runoff, improve uniformity, reduce application rates, and allow a larger area to be irrigated with a given amount of water.

Matched-precipitation rate sprinklers. These sprinklers aid in water conservation by allowing more uniform coverage of a system.

Check valves. Check valves prevent drainage of the pipeline through the low heads, whether integrated in the sprinkler or the lateral line. This saves the amount of water that would be lost from the pipelines each time the system is used, and it also prevents excessively wet areas around low heads. Containing water in the pipelines prevents the introduction of air into lines and minimizes breakage due to water hammer and surge problems. Water hammer and surge can cause a loss of water and damage the landscape.

Master valve. A master valve is an automatic valve in the main line of the system upstream from all automatic valves. It is opened any time the system is operated and is closed when the system completes its cycle. If there is a break in the main downstream of this valve or if an automatic valve fails in the open or running position, the master valve will prevent water from discharging through the break or failed valve, except when the system is within a running cycle. This feature is used in lieu of an excess flow-sensing device. The valve will not signal a problem nor will it prevent loss of water through the failure, but it will limit the loss of water and potential damage.

Wind-sensing devices. An overhead irrigation system will not be useful when wind velocity exceeds 10 mph. An irrigation system, at best, can only function as an overhead flood system under such conditions. Irrigation is usually, or should be, suspended when wind velocities reach 10 mph. Under these conditions, rather than trust suspension of the system to memory, there are features on some equipment that will cause suspension of the watering cycle. The feature may cause resumption of the cycle when the wind tapers off or it may reset to the beginning of the next irrigation cycle.

Rain-sensing device. Automatic irrigation systems suffer public ridicule when they operate during a rainstorm. A rain-sensing device will put an end to this problem and will also conserve water by turning the irrigation system off when sufficient rainfall eliminates the need for supplemental irrigation. This feature, like the wind-sensing device, can either suspend irrigation at a given point in the cycle or will reset to the beginning of the next cycle. This device is also most efficient when used in conjunction with moisture-sensing equipment.

Multicycling of irrigation stations or valves. This feature of an automatic controller is a must when moisture sensing is used. It also is a worthwhile feature on a system without moisture sensing and can conserve water by reducing runoff, while keeping the moisture level in the root zones closer to the optimum level for longer periods of time than nonmulticycling.

Dual programming. Dual or multiprogramming features of an automatic controller can conserve water by enabling the designer and operator of the system to match more closely the water needs of the landscape. When an automatic controller has more than one program available, turf can be watered on a separate and more frequent program than ground cover, shrubs, or trees. This can be accomplished by circuitry which uses one "common" circuit or two individual "commons." Usually one common circuit limits the amount of stations that can be placed on a second program, whereas the two-common circuit approach allows more variations in programming. Another approach for the multiprogramming is the use of electronic or hydraulic pressure signals to allow operation of valves on varying programs.

Electronic accuracy. Mechanical-electrical controllers have always lacked accuracy because their motor and gear wear and cam and level wear affect the unit's timing. Solid-state electronics are now available in the irrigation industry, so timing accuracy is now improved. If minutes of watering are required, the controller will now deliver *exact* minutes of water. Mechanical-electrical controllers desiring 6 minutes of watering used to deliver somewhere between 5 and 10 minutes, and were subject to change, depending on wear and the weather. The days of programming a station to run a minimum of 5 minutes due to the nature of the equipment are also in the past. Most electronic controllers can be accurate to 1 minute.

Use of the irrigation system

Even the best designed system is of little value, if improperly used. An irrigation system properly designed and installed must be operated as it was designed; efforts of the first two portions of the designer-installer-user team will be of little avail. Although users may make some fine-tuning changes in the system to make compatible unions with their particular landscapes, they must be careful to make changes that will not substantially alter the original intent of the systems.

Avoiding problems

Avoiding problems in the use of an irrigation system means watching for the following problems and correcting them as soon as they are discovered:

Failure of any equipment. If a sprinkler fails to pop up or rotate or the water distribution from it is improper, it must be corrected or replaced. If a valve or controller fails to operate properly, it must be repaired or replaced. If any special feature malfunctions, that feature must be reestablished. If there is a line break, it must be repaired.

Loss of sprinklers. If a sprinkler is lost due to vandalism or damage, it not only must be replaced, but it must be replaced by a unit that will perform the same function as the original sprinkler.

Misalignment of sprinklers. Misalignment of part-circle or "two-speed" sprinklers, whether by vandals or other means, must be checked and corrected immediately.

An irrigation system is no different than any other piece of equipment. A preventative maintenance program should be set up and followed. If this is done, the system will serve the needs of the landscape for a long time.

Practicum

10 Influence of water on pest activity

PHILLIP F. COLBAUGH

Phillip F. Colbaugh received his Ph.D. at the University of California, Riverside, while conducting research on the influence of drought on diseases of turfgrass. Presently he is associate professor, urban plant pathology, at the Texas Agricultural Experiment Station in Dallas.

CLYDE L. ELMORE

Clyde L. Elmore received his Ph.D. in botany weed science from the University of California, Davis. As a Cooperative Extension weed scientist, he has been involved for 20 years in turfgrass weed research at UC Davis.

Invasive weed populations and injuries caused by disease and insect infestations are important problems wherever turfgrasses are grown. Annual expenditures for pesticides in professional lawn care markets are increasing rapidly and will approach $190 million by the year 2000.

Harsh environments created by restricted water use may significantly reduce the number of important pests attacking turfgrasses. Damage caused by a few pests capable of adapting to dry environments can be reduced by irrigation practices that limit retention of free moisture in the foliar canopy and upper soil profile.

Water availability is an overriding factor influencing the suitability of habitats and environments for pest activities because water has both direct and indirect influences on activities and persistence of turf pest populations.

Weed, disease, and insect pests are typically a greater problem where water is not a limiting factor for turfgrass growth. Few pests are capable of high damage thresholds in continuously dry environments, and they must rely on temporary periods of available moisture for their damaging activities.

Water is necessary for completion of developmental stages of weed, disease, and insect pests and is often a critical factor affecting their survival from year to year.

Water also affects turfgrass pest populations and activities indirectly by its strong influence on microenvironments where pests survive and initiate damaging activities. Restricted water use for turfgrasses will raise temperatures and lower humidities during the day within poorly developed leaf canopies. These microenvironments are generally not suited for pest activities.

Deficient water limits turfgrass growth and its tolerance to additional stresses caused by pests. In contrast, active growth with adequate supplies of water reduces the impact of pest damage by ensuring recovery from injury through continued vegetative growth.

The distribution and activity of pesticides are frequently influenced by the availability and movement of water in soil and on the turfgrass plant. Uptake and residual activity of pesticides are enhanced in moist environments. Dry environments are not conducive to chemical uptake and residual activity, and toxicity to the turfgrass plant is often increased.

Frequent irrigation during periods of active weed growth favors survival and establishment of turfgrass weeds. Shallow-rooted weeds require more water than deep-rooted weeds.

When turfgrasses are in active growth, high cutting heights and balanced fertility reduce weed establishment and invasion.

The incidence and severity of fungal diseases are largely determined by turfgrass cultural management programs and existing environments favoring pathogenic activities of disease-causing organisms.

Most fungal pathogens have a strict requirement for water and cannot initiate disease during dry periods. Disease activities initiated by spore-forming fungal pathogens are favored in dry environments because of their ease of dissemination and ability to infect the turfgrass plant rapidly.

Saprophytic and parasitic development of fungal pathogens is favored by cyclical patterns of wetting and drying due to an increase in readily available nutrients when dry tissue is remoistened. Desiccation also destroys populations of saprophytic bacteria that normally suppress the growth of fungal pathogens.

Higher cutting heights and irrigation practices that avoid night watering can reduce disease activities by fungal pathogens.

Dry environments reduce developmental stages and feeding activities of insect pests. A few insect and related pests, including chinch bugs and the bermudagrass stunt mite, are capable of damaging turfgrasses under dry growing conditions.

Soil nematodes are dependent on free water in soil for their mobility, feeding, and reproductive activities. Water-conserving practices will greatly reduce populations and activities of soil nematodes.

10 Influence of water on pest activity

Phillip F. Colbaugh and Clyde L. Elmore

Turfgrasses, widely used for their aesthetic and utilitarian values in urban landscapes, also serve important functional roles related to environmental enhancement, such as stabilizing soil, dissipating heat, cleaning air, and reducing noise and glare. Turfgrass acreage is primarily centered in densely populated urban areas. Enormous amounts of money are spent each year on cultural programs to establish and maintain turfgrasses. Labor-intensive cultural practices, specialized equipment, and chemical turf products are often required for turfgrass installation and long-term maintenance programs. By virtue of the acreage involved, the greatest overall expenditure goes to residential lawn care (5). A 1978 estimate of minimum costs to maintain residential lawns in New York was $50 per year (38). Current annual expenditures for maintaining the nation's 68 million residential lawns could easily exceed $3.4 billion. This estimate does not include annual revenues for professional lawn care services ($1.5 billion) or adjusted costs for lawn care products at current prices.

Recent trends have popularized intense culture of home lawns as well as public, commercial, and institutional properties. Rapid growth of the U.S. lawn care industry during the past decade reflects this trend. Increasing cultural intensity has led to emphasis on controlling all types of turfgrass pests, including weeds, fungal diseases, and insects and nematodes. The high regard Americans have shown for pest-free turfgrasses is indicated by annual expenditures for chemical control products. Sales of turfgrass pesticides in professional U.S. lawn-care markets exceeded $155 million in 1979 (8). Projected expenditures for these pesticides in 1988 will approach $190 million.

In an economic sense, major turfgrass pests are those occurring frequently or those that severely reduce the aesthetic or functional values for which turfgrasses are used. The nature of pest injury can, both directly and indirectly, affect the turfgrass plant. Weeds are one of the most common of turfgrass pests because of their ubiquitous nature and obvious presence in uniformly grown stands of turf. Weed populations not only detract from turf's natural beauty, but they also reduce wear characteristics in high-traffic or recreational-play areas. Weeds also directly influence turfgrass vigor and establishment capabilities by competing for available moisture, nutrient supplies, and light. Disease, insect, and nematode pests more directly influence growth characteristics of turfgrasses and their ability to resist environmental and cultural stress injuries. The rapid, destructive activities of these pests occur with little warning and are potential problems throughout the year.

Turfgrass pest problems do not always represent simple damage by one pest population. More than one turfgrass pest can become involved, as when plant injuries caused by the feeding of nematodes or insect pests create entries for plant pathogens through wounds or increase the susceptibility of plants to infection. Weeds frequently invade this weakened turf. Weed populations can similarly serve as a source of pathogen inoculum or as a harbor for insects and related pests capable of reducing the functional and aesthetic value of turfgrasses.

Public awareness of pest problems largely depends on thresholds of damage for individual pests. A single weed may be viewed as a serious pest because of its unsightly appearance; however, an infected leaf producing large numbers of infective spores of a turfgrass pathogen is not readily visualized; thus, its importance is minimal until large-scale disease activity is observed. Forms of disease, insect, and weed problems also vary with the extent or visibility of damage they cause. Disease or insect problems that cause a single dead area are often considered more of a problem than those that cause a general decline or weakening of turfgrass vigor. Populations of broadleaf weeds are easy to detect in a stand of grass; however, high populations of grassy weeds can go unnoticed by the casual observer.

Seasonal activity of turfgrass pests is strongly influenced by environment and turf care. Many weeds, diseases, and insect-related problems occur in warm weather and subside during cooler months. Other pest problems are most severe during fall and winter and occur minimally during summer. Turfgrass management methods often dictate pest control requirements. Such cultural variables as mowing, fertilization, and irrigation are commonly attributed to increased pest populations by their influence on plant growth or through microenvironmental changes they bring about.

Proper cultural practices frequently increase the vigor and ability of turfgrasses to resist or recover from restraining influences of pest populations and their activities. Watering practices are considered to be an overriding factor influencing plant growth and development. Availability of water insures the turfgrass plants' ability to obtain nutrients for continued growth and also helps reduce plant stress. The following discussion identifies important pest problems of turfgrasses and addresses the water-related effects on pest populations and their activities.

Water's influence on turfgrass growth and microclimates

Of all cultural variables used for growing turfgrasses, availability of water probably has the greatest influence on turfgrass performance. Supplemental irrigation is almost always necessary to maintain acceptable growth. Irrigation requirements for specific turfgrasses vary greatly and are influenced by soil characteristics, frequency and duration of natural rainfall, cultural methods, and the rate of water depletion from a stand (65). Conventional irrigation methods employ timed water applications or rely on the appearance of water-stress symptoms to dictate applications. Water retention in low-lying areas and water runoff on high centers or slopes frequently reduce uniform distribution of surface-applied water. Individual turfgrass plants can also be exposed to excessive moisture or to relatively dry conditions, depending on their proximity to the irrigation source. Availability of water on the turfgrass plant or in the underlying soil is thus cyclical and can range from excessive to deficient.

High levels of irrigation and fertilization are frequently used to improve the aesthetics and functions of turfgrasses. Both of these practices can largely be attributed to our inability to supply moisture and nutrients uniformly to millions of individual plants in a typical turfgrass stand. Improved technology aimed at uniform distribution of water and nutrients could greatly improve efficiency in their use. With current cultural inputs, vigorous growth habits of turfgrasses are generally resistant to stress caused by extreme environments and to injuries caused by turfgrass pests. In addition, rapid turfgrass-growing characteristics offer a high degree of flexibility in cultural management programs. Cultivation, thatch removal, close mowing, and other cultural practices injurious to turfgrasses are counterbalanced by abundant supplies of water and nutrients. Use of water-conserving cultural practices is likely to lower significantly the durability and recuperative ability of common turfgrass cultivars.

Populations and activities of turfgrass pests are strongly influenced by microclimate conditions within the turfgrass community. Seasonal activities of many recurring turfgrass pests are often determined by the frequency and duration of moisture supplied during the year. Limitations in water availability or poor distribution of irrigation water can significantly influence turfgrass microclimates. Field measurements of temperatures and relative humidities were compared in moisture-stressed and well-irrigated areas of Kentucky bluegrass (Sydsport) in Santa Ana, California (University of California, South Coast Field Station) following a 1-hour morning irrigation during July (12). Temperature and relative humidity measurements were recorded during the day within the turfgrass canopy of each area. Temperatures within the foliar canopy of moisture-stressed areas of turf (41° to 51°C) were considerably higher than temperatures (32° to 40°C) recorded in well-irrigated areas (fig. 1). The highest temperature recorded in areas receiving adequate irrigation (40°C) was almost 11°C lower than high temperatures recorded in thinly developed dry areas. Measurements of relative humidity within the two areas of bluegrass also indicated the presence of harsh environments within moisture-stressed turf (fig. 2). Relative hu-

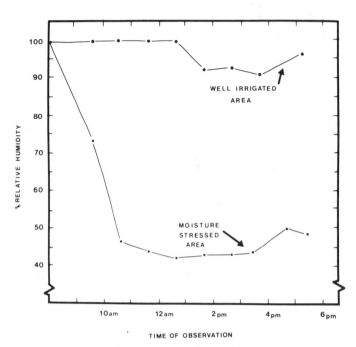

FIGURE 2. Relative humidities within well-irrigated and moisture-stressed areas of Kentucky bluegrass following a 1-hour morning irrigation (8 a.m. to 6 p.m.), Santa Ana, California

midities within the leaf canopy of turfed areas receiving adequate irrigation varied slightly from 100 to 91 percent throughout the day. In contrast, humidity levels within moisture-stressed areas dropped rapidly only 2 hours following irrigation where they remained at 45 percent for most of the day.

Depending on cultural and climatic conditions where water use is restricted, microenvironmental effects of water conservation may not be as pronounced as the example used in this study. Contrasting microenvironments developed in well-irrigated and moisture-stressed areas of turfgrass do, however, show the influence of water as an important regulatory factor affecting turfgrass microclimates.

Turfgrass weeds

A weed was defined by Ralph Waldo Emerson as "a plant whose virtues have not yet been discovered." In monocultures of turfgrasses, broadleaf weeds are easily observed and are usually considered objectionable. Grass species, if not greatly different in color, texture, or growth habit (rate of growth or prostrate habit) have not been as objectionable. However, if the turfgrass is subjected to such stresses as wear from traffic, compaction, overwatering, or drought, and the characteristics of the various species of plants differ, then the competitive edge is given to a single species and it dominates. Often this dominant species is not the "desirable" turfgrass.

Weeds are a natural component of the landscape. In some areas a species may be favored as the adapted turfgrass, whereas nearby or in other areas it is considered weedy. This is especially true with such species as common bermudagrass (*Cynodon dac-*

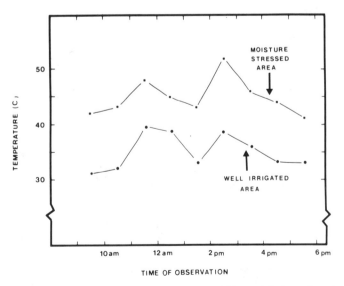

FIGURE 1. Leaf canopy temperatures within well-irrigated and moisture-stressed areas of Kentucky bluegrass following a 1-hour morning irrigation (8 a.m. to 6 p.m.), Santa Ana, California

tylon [L.] Pers.), tall fescue (*Festuca arundinacea* Schreb.) or even annual bluegrass (*Poa annua* L.). In minimal maintenance areas, any plant that survives and becomes dominant may be desirable and may not be considered a weed.

Hybrid varieties are frequently produced (bermudagrasses) or naturally selected biotypes (annual bluegrasses) that are used as turf species in local areas. Some plants can also be desirable plants, although they are normally thought of as weeds if they are utilized to fill in weak spots in a turf or as a transitional species during the establishment of desirable turfgrasses.

More than 50 plant species are considered as weeds in turfgrasses (22, 30). Some 20 species are commonly found in various regions of the U.S. (23). These species have been distributed by human beings and their animals and by their movements across the United States. Plants as weeds are categorized into broadleaf weeds (dicotyledonae or "dicots") and narrowleaf weeds (monocotyledonae or "monocots"). Sedges (Cyperaceae) are grouped within monocotyledonae. Narrow-leaf weeds are principally grasses, although wild onion and wild garlic (*Allium* spp.) are in the same group. These plants are further divided by life-cycle characteristics. Annual weeds are characterized as those germinating from seeds, progressing through a vegetative stage, flowering, and maturing seed within a year. Where there are distinct seasons, annuals are frequently found to occur either as winter- or summer-season plants. Summer annual weeds germinate in spring, grow during summer, and mature before a frost. Crabgrass (*Digitaria* sp.) and knotweed (*Polygonum aviculare* L.) are examples. Conversely, winter annual weeds germinate in fall, grow throughout winter, and mature in spring before the heat of summer. Henbit (*Lambium amplexicaule* L.) and chickweed (*Stellaria media* [L.] Vill.) are examples.

Biennial plants have life cycles that encompass 2 years. Seeds germinate the first year and form a rosette. The second year, after a temperature vernalization or a day-length reaction, the plants will form a flower stock, set seed, and die. Examples include bristly oxtongue (*Picris echiodes* L.) or cheeseweed (*Malva parviflora* L.).

Perennial weeds can live indefinitely, although the tops may die during winter. Once started, perennial plants remain until they are killed. They may start from germinating seeds, but many plants spread naturally by vegetative means such as root pieces, rhizomes, stolons, or tubers. Once established in turf areas, perennials spread by competitive encroachment of vegetative growth and/or by seed.

Plants often coexist in turf areas until natural or artificially induced stress is applied. The planting and maintaining of turfgrass monocultures could be considered a stress to a planting site. Residual plant species from introduced seed or vegetative propagules in the site do not just disappear. By changing watering regimes (moist or dry) stress is applied to the existing plants, often resulting in population shifts to adapted plant species. Plants may appear in turf as weeds under the following water regimes:

- During temporary or extended periods of drought;
- When excessive water is present;
- When frequent, light irrigations are used to keep the surface continually moist; or
- When rainfall or irrigation occurs frequently during heavy weed-seed germination (fall and spring).

Certain plant species are frequently associated with climatic (temperature and water) areas. Modification by irrigation has introduced or reduced some plant species. Weed species and their characteristics will be discussed when they have been reported (1, 2, 6, 30, 39, 59, 61, 67) or observed to favor a water regime.

During germination, weed seeds need a continuous supply of moisture. If a seed germinates in thatch or on surface soil, frequent irrigation or rainfall will allow the seedlings to establish. If however, the thatch and soil surface are allowed to dry, the seedlings will die. Weed seedlings are encouraged by light, frequent irrigation.

Large crabgrass (*Digitaria sanguinalis* [L.] Scop.) and **smooth crabgrass** (*Digitaria ischaemum* [Schreb.] Muhl.) (30, 61, 67) are summer annual grasses that spread by seed. Although crabgrass may germinate throughout summer, spring germination starts in the presence of light when the soil temperature reaches 16°C (61). Seedlings are susceptible to drought at germination; thus, frequent watering in spring enhances establishment. Alternate drying and wetting of the soil is a control measure. Large crabgrass germinates approximately 2 to 3 weeks before smooth crabgrass does in California. Once established, crabgrass spreads by rooting at the nodes of prostrate stems. Fall frosts kill crabgrass. In protected sites in the southern United States crabgrass lives into the second year.

Goosegrass (*Eleusine indica* [L.] Goertn.) is a summer annual that requires moist conditions for seed germination, but once established it is tolerant of drought, compacted soils, and traffic. Goosegrass germinates much later than crabgrass. It is spread principally by seed, but the stems will sometimes root at the node.

Dallisgrass (*Paspalum dilatatum* Poir.) is a clump-forming perennial grass that spreads by seed and rhizomes. Its seeds are flat and pubescent (hairy); thus, surface-flowing water will help its spread. Germination is favored by frequent irrigation. Once established, it will tolerate drought, but it flourishes in moist and, at times, saturated conditions.

Creeping woodsorrel (*Oxalis corniculata* L.) and **yellow woodsorrel** (*O. stricta* L.) are broadleaf plants. Creeping woodsorrel is a perennial with a heavy taproot and creeping, rooting stems; *O. stricta* is an annual. Germination and establishment requires frequent light irrigation or rainfall to keep the surface moist in the absolute presence of light (40). Overwatering of turfgrass contributes to problems with *Oxalis* spp. (2).

Annual bluegrass, often referred to as *Poa annua*, is a winter annual in warm climates and a summer annual in northern regions. Seeds germinate rapidly at 10° to 15°C, and vegetative growth is favored at temperatures of 16° to 21°C. *Poa annua* requires high levels of moisture to germinate and to get established. Stress from drought and high temperatures will reduce vigor or kill seedlings and established plants. It tolerates compacted, wet soils. It is very susceptible to heat stress, especially under dry conditions. Nearly all annual bluegrass plants are killed in regions where summer temperatures are high. Annual bluegrass is susceptible to common turfgrass diseases, especially during warm weather. *Poa annua* invasion would be expected to be reduced and controlled more easily by water conservation practices where high temperatures prevail. Syringing of turfgrasses helps to maintain an environment for germination and growth of annual bluegrass (1).

Rough bluegrass (*Poa trivialis* L.), a perennial cool-season plant, tolerates shade and looks like *Poa annua*. Rough bluegrass (rough-stalked meadowgrass) is sometimes used as a grass for overseeding, but it frequently remains as a weed. It requires wet soils and frequent irrigations in summer to keep it from going dormant and turning a yellow-to-brown color. The crowns revive in cool wet periods in fall.

Common chickweed (*Stellaria media* L.), a widely occurring, creeping winter annual, is common in open patches or thinned turf. It reproduces by seed and creeping stems that root at nodes in contact with the soil. Seed germination during fall is favored by cool, moist environments. Usually presence of the weed indicates poor turfgrass density during much of the year. Improvements in turfgrass cover, supplements of fertility, proper mowing, and judicious watering are recommended to control the problem. Common chickweed grows best in cool weather or in winter in sheltered areas.

Mouseear chickweed (*Cerastium vulgatum* L.), a widely distributed prostrate perennial, forms dense patches and typically invades open patches or even established turf. It spreads principally by seed, but spreads in patches by rooting at nodes. Distribution is primarily limited to cooler regions of the U.S. *Cerastium vulgatum* requires cool, moist habits. Hot, dry climates often destroy it.

Nutsedges (*Cyperus esculentus* L., *C. rotundus* L.), perennial sedges spreading by seeds and tubers, prefer moist or wet soils. Once established, plants will tolerate extended drought conditions. The weeds are commonly referred to as nutgrass. Nutsedges tend to grow rapidly, and commonly the leaves are taller than the intended turfgrass species. Underground tubers are formed and may remain dormant for long periods.

Bentgrass (*Agrostis palustris* Huds.) is a stoloniferous species that spreads rapidly in cool moist areas. Although it is used extensively for golf course greens, it escapes and is frequently found as a weed in cool-season turf. It is spread vegetatively and by seed.

Influence of cultivar and cultural care. A turfgrass species well adapted to any particular site is best suited to the establishment of a fine turf area. Also, this adapted vigorous species is less prone to weed invasion. The competitive characteristics of the turfgrass, such as better utilization than the weed seedling of water, nutrients, or light, favors the turf. These same characteristics may make the grass (bermudagrass) a weedy plant where plant-type zones (transitional zones) overlap. Different varieties of *Lolium perenne* L. were found to vary in their capability to reduce weed invasions or, conversely, to allow weeds to establish (36). Pennfine, Derby, Clipper, Diplomat, and Manhattan, for example, had only 21.5 percent or less weeds, whereas Yorktown and Ensporta were 35 and 45 percent weeds, respectively.

Each turfgrass species has an optimum and mowing tolerance range that varies by species, climate, culture, and other environmental influences. The optimum cutting height for the species at any site will give optimum protection against weeds. The greater top growth of 'Merion' Kentucky bluegrass turf has been related to greater depth of rooting and, also, to greater water loss (45). The low cutting height of cool-season turf species has been found to be a major factor favoring *Poa annua* increase on golf fairways (30). On the other hand, high mowing (3 inches) discourages annual bluegrass invasion into bermudagrass compared with a ½-inch mowing height (69). A mowing height and weed invasiveness study, using perennial ryegrasses, showed that 11 of 12 varieties had significantly fewer weeds when mowed at 1½ inches than at ¾ inch (36). Although there may be greater water loss from high-mown turf (45, 54), there is a distinct reduction of weed invasion.

The greatest influence of fertilization on weeds is the indirect effect. Fertility increases grass vigor, thus reducing the competitiveness of the weeds. Nitrogen is the limiting element in most areas. Applications of phosphorus or more frequently lime (acid soils) will give a growth response in turfgrasses. Not only is the amount of supplemental elements critical, but so is the application's timing. If fertilizers are applied to turf at the optimum time to insure vigorous grass growth, weeds cannot invade and compete. Fertilizer applications should be made to enhance turf before crabgrass germination in spring or annual bluegrass in fall or spring, depending upon locality. A well-fertilized, adapted turf will reduce invasion by dandelion (*Taraxacum officinale* Weber), large crabgrass, smooth crabgrass, sweet clovers (*Melilotus* sp.), and hop clover (*Medicago* sp.).

Individual turf varieties vary in their responses to nitrogen and in their effectiveness to inhibit weed invasion. In a study of 12 varieties of perennial ryegrass, six species (Common, Diplomat, Ensporta, Lamora, NICS-321, and Wendy) were found to show no significant difference in weed invasion, using fertilizer rates of 2, 4, or 6 lb/1000 sq ft (36). All other varieties (Pennfine, Manhattan, Derby, Clipper, and Yorktown) were much weedier in the 2 lb/1000 sq ft fertilization regime than at the 4 or 6 lb/1000 sq ft rate.

Water's influence on weeds. Weed seeds are frequently moved about with water. Weed dissemination for short distances can occur with rainfall or irrigation. Examples include movement of light seeds, such as dallisgrass and annual bluegrass, with surface water on the soil. These seeds will float and accumulate in low areas or where debris or an obstruction occurs. If irrigation water is drawn from surface waters (ponds or open canals), weed seed is frequently moved with this water through sprinklers or more commonly with flooding irrigation. Weed seeds are dropped onto water from weeds surrounding ponds or along canal areas. These small seeds can be transported through irrigation waters.

Many weed species are favored by frequent irrigation or overirrigation (6, 59, 67, 70). As stated by Engel and Ilnicki (30), "An abundance of water is necessary for weed seeds to germinate and become established when in competition with a mature turf cover." Annual bluegrass is most frequently mentioned as the species that will increase, invade, or sustain itself under this regime. In southern California, annual bluegrass maintained itself in bermudagrass turf in summer with ½-inch irrigation every 3 to 4 days, whereas ¾-inch of irrigation every 10 days did not affect bermudagrass quality, but annual bluegrass declined (68). In a fine-textured soil, annual bluegrass grew well at 50 to 60 percent of soil water-holding capacity (39). When the moisture level was maintained at 30 percent of water-holding capacity, very little annual bluegrass remained. Spraying greens (syringing) to control temperature stress resulted in salt accumulation. By increasing irrigation to control salts, *P. annua* was favored (1). *Oxalis* sp. and annual sedge germination and establishment were enhanced by overwatering. Crabgrass (*D. ischaemum*) or common chickweed was encouraged with continuous dampness.

While establishing *Poa pratensis* L., *Festuca rubra* L., or *Agrostis palustris* turf with sprinkler irrigation in Ohio, increased watering regimes (1.5, 2, and 3 times normal) reportedly increased the number and size of velvetgrass (*Holcus lanatus* L.) infestation, particularly in fescue turf (66). In the same study, clover also increased in fescue turf with increased irrigation.

Watering above the limits of turfgrasses generally favors spread of weedy plants in established turf areas. Plants that root at the nodes, such as crabgrass and common chickweed, spread over the surface in turf. Bermudagrass was more invasive in bentgrass turf under irrigation intervals of 10 days than in daily to 5-day intervals (44). Fescue and bluegrass turf was least invaded by bermudagrass at the 10-day irrigation intervals. The explanation of why bluegrass was not invaded by bermudagrass at the 10-day interval was that the bluegrass must have been "adapted to the water stress."

Influence of irrigation practices on weeds. Irrigation practices that supplement natural rainfall can enhance turfgrass growth. They also frequently alter weeds or the opportunity for weediness in turf. This may be due to excessive irrigation or extended periods of drought stress.

Wilson and Lathan (68) stated: "Annual bluegrass is becoming more of a problem due to excessive nitrogen levels, low cutting, and too much water." Weeds, such as velvetgrass (66, 70) and rough bluegrass (70), are reportedly found much more frequently in areas where soil is kept wet for longer periods. Many weeds (Table 1) have been noted to require high levels of moisture at germination and establishment. If the soil surface and/or the thatch remains moist, the seedlings can establish. If the irrigation or rain cycle is far enough apart to dry the soil surface, the seedlings cannot establish. Once weed seedlings are established, frequent watering is not so critical to maintain their development. Other weeds require relatively high soil-moisture levels throughout their life cycles (Table 2). Stress of low moisture, usually associated with high temperatures, will "melt out" such weeds as annual bluegrass and common chickweed. During these high temperature periods, goosegrass and crabgrass become more prevalent.

The summer annual, crabgrass, is most often found in thin areas of turf or in low areas where soil moisture levels are higher (52). Observations in California also indicate that light, frequent irrigation (keeping the soil moist) in late winter and early spring enhances seed germination and establishment of crabgrass seedlings. If irrigation is light enough to allow the soil surface to dry, this regime does not enhance establishment of crabgrass. The most critical factor for weed establishment is maintaining a moist turf. Greenhouse studies show that in soil kept moist (two to three turfs per week to field capacity) more crabgrass became established than in drier turf (daily wetting of the top 1/2 to 3/4 inch of soil only), even if herbicides are used for control (52).

In an evaluation of the invasion of common bermudagrass into tall fescue, bentgrass, and bluegrass turf, with a daily or once-in-5-days irrigation, bermudagrass invasion was much more rapid in tall fescue and bluegrass than when the turf was irrigated at 10-day intervals (44). Bentgrass was invaded most rapidly by bermudagrass when irrigated at 10-day intervals, indicating a higher requirement of shallow-rooted bentgrass for water to be competitive.

When a manual turf treatment to simulate treatment by local turfgrass managers was compared in bermudagrass turf to four other irrigation criteria, weeds were found to be much more prevalent in the manual irrigation area (Table 3) compared with 15, 40, or 65 cb, as measured by tensiometers in *Cynodon dac-*

TABLE 1 Weed species requiring high soil moisture during germination

Annual	
Large crabgrass	*Digitaria sanguinalis*
Smooth crabgrass	*Digitaria ischaemum*
Goosegrass	*Eleusine indica*
Yellow woodsorrel	*Oxalis stricta*
Perennial	
Dallisgrass	*Paspalum dilatatum*
Rough bluegrass	*Poa trivialis*
Creeping woodsorrel	*Oxalis corniculata*

TABLE 2 Weed species requiring high soil moisture throughout life cycle

Annual	
Annual bluegrass	*Poa annua*
Common chickweed	*Stellaria media*
Perennial	
Bentgrass	*Agrostis palustris*
Dallisgrass	*Paspalum dilatatum*
Mouseear chickweed	*Cerastium vulgatum*
Yellow nutsedge	*Cyperus esculentus*
Velvetgrass	*Holcus lanatus*

TABLE 3 Effect of five irrigation regimes on *Poa annua* in *Cynodon dactylon* turf

Irrigation regimes	*Poa annua* in *Cynodon dactylon* (plants per plot)
Automatic @ 15 cb	0.8 z*
Automatic @ 40 cb (35 for cool)	0.0 z
Automatic @ 65 cb (55 for cool)	0.0 z
Manual—75% (cool period) to 87% (summer ET)	4.8 z
Manual (simulate local managers)	15.2 y

SOURCE: Youngner et al. (70)

*Numbers in the same column with the same letter not significantly different P = 0.05% (Duncan's Multiple Range Test).

tylon turf (70). When the same study was conducted in *Festuca arundinacea* and *Poa pratensis* turf, there was no significant difference in *P. annua* invasion in *P. pratensis* turf. Irrigation at 35 and 55 cb had significantly less *P. annua* per plot than 15 cb or evapotranspiration (ET) regulated irrigations. Manual irrigation to simulate that by local managers watering had significantly more *P. annua* in *F. arundinacea* turf.

Effects of minimum water. No study was found to indicate the relative amount of water utilized by a weedy turf, as compared to a well-maintained turf. Weeds frequently tolerate drought or low-moisture levels to a greater extent than do turfgrasses (Table 4). Wilson (67) measured the water-supplying capacity of many turf sites around Baltimore, Maryland. White clover (*Trifolium repens* L.) and narrow-leaf (buckhorn) plantain (*Plantago lanceolata* L.) were found to be very drought tolerant. Bluegrass turf responded more rapidly to drought-stress injury as soil moisture declined than did white clover. When rain was received, however, bluegrass revived more quickly in fall. During summer bluegrass did not revive. Red fescue (*Festuca rubra* L.) was judged more drought tolerant than bluegrass. Perennial forms of *Poa annua* were found to be dominant in areas of low-moisture stress, whereas the prolific, seed-producing upright annual form was found in moisture-stressed areas (6, 34).

Minimal water supplies in warm climates or transitional zones stress marginally adapted species and encourage weed invasion. In Davis, California, a transitional area for cool-season grasses, bentgrass watered at 10-day intervals was invaded by bermudagrass. Irrigated daily or at 5-day intervals bentgrass suffered fewer invasions (44).

TABLE 4   Weed species in turfgrass tolerant of low soil moisture

Annual	
Knotweed	*Polygonum aviculare*
Goosegrass	*Eleusine indica*
Russian thistle	*Salsola kali*
Kochia	*Kochia scoparia*
Perennial	
Buckhorn plantain	*Plantago lanceolata*
Bermudagrass	*Cynodon dactylon*
Dandelion	*Taraxacum officianale*
Yarrow	*Achillea millefolium*
Field bindweed	*Convolvulus arvensis*
White clover	*Trifolium repens*

Effects of irrigation on herbicide activity. Herbicides are used extensively on turfgrasses for selective control of grass and broadleaf weeds. Preemergence herbicides are applied in fall or late winter to early spring, depending on the species to be controlled. These materials control germinating seedlings before they can become established. Postemergence herbicides are applied to the foliage of established weeds. They either suppress or kill on contact (contact herbicides) or move from the point of application on leaves or stems to a site where a critical metabolic process is affected (translocated or systemic herbicide).

Preemergence herbicide activity is often affected by the amount and timeliness of irrigation. Some herbicides were found to require only ¼ to ½ inch of water initially for their activation (43). With these products, 2 inches of rainfall with the first irrigation leaches the herbicide too deeply into the soil to give acceptable weed control. If a light rain or irrigation occurs and is then followed by additional rainfall, the amount of subsequent moisture is not so critical in the movement of the herbicide into the soil. An herbicide that exhibits these tendencies is pronamide (*Kerb*). Also, if herbicides are applied to dry soils, there is less movement with subsequent water than when materials are applied to wet soil. Other herbicides, such as DCPA, benefin, oxadiazon, or pendimethalin adsorb so tightly on soil colloids (organic matter and clay) that leaching is not a concern.

Sandy (sand, sandy loam, or loamy sands) soils, because of their low adsorbing characteristics, can greatly influence herbicide activity in soil. With increased use of sand-based greens, more concerns have been raised about the effectiveness of pre-emergence herbicides. Generally, the residual effectiveness of an herbicide like bensulide will be shortened on sand greens compared with heavier soils.

An area with virtually no information available but of real concern is the effect of soil-wetting agents on herbicide movement in soil. In an evaluation of the leaching of diuron alone or in combination with different adjuvants, it was found that movement in soil varied greatly with different mixtures (4). Diuron, mixed with adjuvant A (X-77), followed by 10 inches of water, would move 12 inches in the soil. With adjuvant B, diuron did not move from the surface. Diuron alone moved 6 inches with 10 inches of water. Thus, some wetting agents apparently drastically affect the leaching or adsorbing characteristics of some herbicides. Information is not available for commonly used turf herbicides.

Postemergence herbicides can be affected in two ways. First, the herbicides DSMA, MSMS, and phenoxy (2,4-D, mecoprop, dichlorprop), and anisic acid materials (dicamba) are affected adversely by irrigation or rainfall immediately following application. Not only are the herbicides' effectiveness reduced, but root uptake of 2,4-D and other phenoxy materials can occur and cause grass injury. The soil activity of dicamba through the roots of trees and shrubs can be increased with increased irrigation.

Fungal diseases

While diseases are often defined to include both biotic and abiotic causes, most diseases of turfgrasses are caused by infectious agents that produce abnormalities in plant growth. Widely occurring turfgrass diseases are caused by pathogenic fungi and soilborne nematodes (18, 64). For purposes of distinction, nematode-incited diseases will be discussed separately.

Most infectious diseases of turfgrasses are caused by fungal pathogens that enter the plant and produce disease symptoms. Whenever temperatures and moisture conditions are favorable and parasitic fungi are present, fungal diseases can develop on susceptible turfgrass cultivars. Fungal pathogens occur as microscopically thin filaments or mycelia that are parasitic on the turfgrass plant. At times, these filaments can be seen as a cobwebby growth on the surface of infected plants. Some fungal pathogens are obligate parasites that obtain food only from living plant parts; others are facultative parasites that obtain food from organic residue or from living plants. With the exception of the rust and powdery mildew fungi, all other fungal pathogens of turfgrasses are facultative parasites.

Pathogenic activity is initiated with spore germination or the growth of surface mycelia into the living plant. Entry of fungus mycelia into the plant is accomplished through wounds, stomates, or directly through cell walls by the production of cell wall degrading enzymes. Upon entering the host, fungal pathogens can obtain nutrients from both dead and living plant cells. Obligate parasites, including the rust and powdery mildew fungi, cause little damage to host cells and obtain their food by diverting nutrients from the host to the infecting mycelia. Facultative parasites are more destructive in their pathogenic activity and obtain food from dead or dying plant cells killed by the action of enzymes or toxins released from the infecting mycelia. Following initiation of disease activity, spores are commonly produced on the surface of diseased plant parts. Spores or mycelia produced by the infecting pathogen can infect the same or adjacent plant parts or are carried to distant plants by the action of wind and water, mechanically by mowing equipment, or by infected plant parts such as grass clippings.

The rapid development of fungal diseases by production of spores or growth of mycelia distinguishes them from other problem pests of turfgrasses. Environmental conditions favoring disease activity can result in severe losses of turfgrass in a short time. Diseases are known to occur annually or during particular seasons of the year; however, their incidence and severity can vary greatly from year to year (24). Environmental relationships and management practices play major roles in governing the extent of disease activity on turfgrasses. In general, cultural programs designed for the production of actively growing turf can reduce the incidence and severity of fungal disease. Avoiding plant stress and maintaining adequate water balances on turfgrasses, as well as supplying balanced nutritional programs, have proved helpful. Modification of cultural practices with respect to irrigation, fertilization, soil aeration, drainage, and soil pH, as well as changes in mowing height, frequency, and removal of thatch, are useful for limiting fungal disease (18, 58, 64).

Practical experience has shown that sound turfgrass management programs do not always prevent disease. Severe forms of disease often occur on highly sensitive cultivars or during environmental periods favoring pathogen development, regardless of management programs employed. Strategies for use of disease-resistant cultivars are also limited because of the potential fungal diseases that can occur during the year. Resistance of a variety to one disease may be complicated by a higher susceptibility to other diseases. Outbreaks of turfgrass diseases on disease-sensitive cultivars or during environmental periods encouraging disease activity often rely on chemical controls capable of contending with higher levels of disease pressure.

Seed decay and **damping-off** affect newly planted seed used to establish lawns or temporarily overseeded grasses grown in the South. Seedling diseases can be caused by one of several soil or seedborne fungal pathogens; however, severe seedling disease activity is frequently associated with *Pythium* and *Rhizoctonia* spp. under moisture-saturated growing conditions.

When seed germination and seedling growth are restricted by suboptimal growing conditions, seedlings become increasingly susceptible to seedling diseases (58). Reduced seedling growth, caused by excessively moist conditions or restrictive growing temperatures, is often ideally suited for pathogen activity. Timing in the planting of seed should insure optimal moisture and temperature conditions for rapid seedling establishment. Excess moisture can be reduced by increasing air circulation, improving drainage characteristics of seedbeds, and using minimal amounts of supplemental watering during the establishment of seedling stands (31, 32).

The **Helminthosporium diseases** occur widely and damage almost all turfgrass species (58). Causal agents include *Drechslera*, *Bipolaris*, and *Exserohilum* spp., which are separated into more than 20 distinct fungal species with pathogenic activity on turfgrasses. Leaf spotting and foliar blighting are initiated by spores produced on decomposing organic debris surrounding the turfgrass plant. Spores arising from the debris layer are spread to healthy plant parts by wind, splashing water, mowing operations, and disturbances created by vehicles and foot traffic. Irrigation practices frequently determine the incidence and severity of Helminthosporium diseases on turfgrasses. Alternating patterns of wetting and drying favor production of spores on turfgrass debris. In contrast, extended periods of leaf wetness favor infection by spores deposited on turfgrass foliage. Moist conditions during rainy weather or following irrigation at night often result in severe disease activity.

Brown patch, an important disease on most turfgrass species grown in the United States, primarily occurs during warm, humid weather and can spread very rapidly. Symptoms of foliar blighting cause circular areas of dead grass, ranging from a few inches to 50 feet or more in diameter. *Rhizoctonia solani* Kuehn is the primary causal agent of brown patch; however, a related disease caused by *R. cerealis* van der Hoeven is active during cooler periods and is referred to as cool-season brown patch (9). Both pathogens produce disease symptoms by growth of infectious fungal mycelia. In the northern states and Atlantic coast, warm-season brown patch can occur throughout summer. In other areas, the disease is usually greatest during the transition seasons of spring and fall because high temperatures and dry environments during hot summers restrict disease activity by *Rhizoctonia* spp.

Dollar spot, caused by *Sclerotinia homeocarpa* F. T. Bennett, occurs throughout the United States in late spring, early summer, and fall, during moist periods with warm days and cool nights. Prolonged periods of high humidity are necessary for mycelial growth activity by fungi causing the disease. Small spots of diseased turf appear as sunken areas of dead foliage; under severe conditions the spots can coalesce to destroy large areas of turf. Dollar spot occurs on all the major turfgrasses and is particularly severe when high or low nitrogen fertility is coupled with moisture stress (18, 64). Bentgrasses used for greens of golf courses are highly susceptible. In the South, bermudagrasses and zoysias are primarily affected during early summer and fall season.

Pythium blight is sometimes referred to as greasy spot or cottony blight. Root and foliar diseases caused by *Pythium* spp. are particularly severe on bentgrass and ryegrass; however, foliar blighting can occur on almost any turfgrass in a soft or succulent state of growth (31, 53). Excessively moist conditions are required for severe disease symptoms and spread of Pythium blight. Affected turf becomes spotted with blighted areas ranging from 1/2 inch to several feet in diameter. In the disease's early stages affected grass leaves appear water soaked. The entire shoot then collapses, and the affected areas fade from green to light brown. Pythium blight is most prevalent during rainy weather or in low-lying areas where air circulation is restricted. Under ideal conditions for disease activity, *Pythium* spp. are capable of spreading very rapidly over large areas of turf.

Fusarium blight, caused by *Fusarium culmorum* (W. G. Smith) Sacc. and *F. tricinctum* (Corda) Sacc., annually destroys large acreages of Kentucky bluegrass on home lawns and on golf courses (57, 64). Symptoms include foliar blighting and a crown and root rot. The disease appears as circular dead spots ranging from 1 to 2 feet in diameter. Its severity is greatly increased by high levels of nitrogen fertility and moisture stress. The greatest activity commonly occurs during hot summer months when moisture stress weakens turfgrass vigor. Watering practices should insure water penetration deep within the soil profile to avoid water deficiencies during the hot summer months. Light and frequent water applications during summer help reduce midday heat stress. Kentucky bluegrass cultivars are available with resistance to Fusarium blight (64). Mixing perennial ryegrass seed with bluegrass at the time of planting has also reportedly suppressed Fusarium blight disease activity successfully (35).

Rust diseases caused by *Puccinia* and *Uromyces* spp. occur on all turfgrasses, affecting large areas. They can occur at any time in regions with mild winter climates or during most of the growing season where climates are more severe. Affected turf appears unthrifty, and the grass becomes chlorotic as if suffering from a nitrogen deficiency. Repeating spore stages of the rust fungi are produced on turfgrass leaves and leaf sheaths. The rust diseases are often severe on turfgrasses grown in stressful environments. Several turfgrass cultivars are known to resist them; however, new races of rust fungi potentially threaten resistant varieties.

Nearly 50 species of soil basidiomycetes are associated with the **fairy ring diseases** (18). Usually the first noticeable symptom is the appearance of a circular or semicircular band of mushrooms or chlorotic streaks in stands of turfgrass. Mushrooms are fruiting structures of fungi causing fairy rings and are primarily produced during moist periods in spring and autumn. As the mushrooms die, nutrients are released that can cause grass to become greener within the ring area. Grass at the edge of the rings may turn chlorotic because of parasitic activities of the fungus as it grows in the outer edge of the diseased pattern. Fairy rings are commonly observed in locations where tree roots or other large plant parts are decomposing in the underlying soil. Fungi that cause fairy rings are also capable of growing on decomposing plant debris comprising the thatch layer of turf.

Stripe smut, caused by *Ustilago striformis* (Westend.) Niessl., attacks Kentucky bluegrass (64). Smut is systemic; once plants become infected, they remain so for life. A smut-infected plant is always weaker than a healthy plant and, consequently, any additional stress usually results in death (64). The most common form of environmental stress, drought, usually kills entire lawns where summer irrigation is not practiced regularly. Smut-infected plants do not tiller as profusely as healthy plants; consequently, bare areas are not readily filled, resulting in weed invasions. Stripe smut symptoms usually appear as yellow blades of grass in affected turf. As the disease advances, leaf blades curl and have black stripes running parallel the length of the blades. When infected blades are disturbed, there is observed a black sootlike dust, the liberation of many fungus spores. Stripe smut symptoms most commonly occur in spring and fall during cool, moist weather when day temperatures are below 70°F. The fungus gradually disappears as temperatures become warmer. Most of the infected turf is lost during the hot, dry weather of summer, when the grass is under heat and drought stress, or during winter, when plants are subjected to desiccation and freezing temperatures.

Gray leaf spot, caused by *Pyricularia grisea* (Cook) Sacc., primarily affects St. Augustinegrasses grown in the South. The disease causes circular or oblong leaf spots that are brown to ash gray with brown to yellow margins. Severely infected leaf blades wither and turn brown. Spores are produced in large numbers on infected leaves. The disease is usually more damaging in shaded areas where dew remains on the grass for long periods. Excessive levels of nitrogen fertility tend to enhance disease development.

Water influence on fungal diseases. Water is required for development of all fungal organisms. Few fungi can maintain active growth and development at relative humidities below 98 percent (11). For most fungal pathogens, moisture-saturated air or free water is required for all stages of growth and development. This includes germination of spores (fig. 3), growth of mycelia, infection of the turfgrass plant, and production of spore inoculum. Because all of these activities require water, the amount of fungal development and severity of disease will increase as water availability increases. Watering practices and the occurrence of natural rainfall thus influence the incidence and severity of fungal diseases.

The frequency and duration of water supplied during the year often determine the suitability of habitats for development

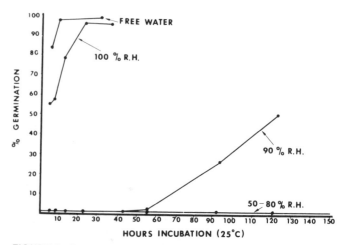

FIGURE 3. Percentage germination of spores of *Bipolaris sorokiniana* maintained at varying relative humidities over a 150-hour period at 25°C (12)

of fungal parasites. Lengthy periods of supplemental irrigation or rainfall encourage the presence of humid microenvironments and retention of free moisture on the turfgrass plant. These environmental conditions, coupled with moderate temperatures, favor development of most fungal diseases. The formation of dew and release of guttation fluids on turfgrass foliage also favor the occurrence of fungal diseases (25, 29). Both dew and guttation fluid are formed in the presence of moisture-saturated environments (62). Timing of irrigation can be restricted to morning hours to reduce leaf wetness and high humidities favoring turfgrass disease activity. Irrigation at night frequently increases activity due to longer periods of moisture retention within the foliar canopy. Long periods of excessive moisture favor development of several important fungal diseases, including dollar spot, gray leaf spot, Pythium blight, Rhizoctonia brown patch, and seedling diseases (Table 5). While these diseases are increased by excessively moist environments, the nutritional status of turfgrasses can also influence severe disease activity (18, 58, 64).

TABLE 5 Relative moisture conditions favoring the incidence and severity of common fungal diseases of turfgrasses

Diseases favored by continuously moist environments	
Gray leaf spot	*Pyricularia grisea*
Pythium blight	*Pythium* spp.
Rhizoctonia brown patch	*Rhizoctonia* spp.
Seedling diseases	*Rhizoctonia* and *Pythium* spp.
Diseases favored by wetting-drying cycles or low soil moisture	
Dollar spot*	*Sclerotinia homeocarpa*
Fairy ring diseases	Soil basidiomycetes
Fusarium blight	*Fusarium culmorum, F. tricinctum*
Helminthosporium diseases	*Drechslera* and *Bipolaris* spp.
Rust diseases	*Puccinia* and *Uromyces* spp.

*Severe disease activity favored by low soil moisture and periods of free moisture on the foliar canopy.

Splashing water and free-flowing water help disperse pathogen inoculum. Spores and other forms of inoculum are commonly spread from decomposing organic debris or infected foliage by splashing water during irrigation cycles or periods of rainfall. Irrigation water and natural rainfall can also effectively redistribute inoculum on the turfgrass plant. Spores of *Drechslera* and *Bipolaris* spp. are often washed from leaf blades to critical areas of infection on turfgrass crowns or culms by the action of surface-applied water. Free-flowing water can disperse infected plant debris or pathogen inoculum for long distances. The movement of groundwater particularly influences the rapid spread of Pythium diseases by dispersal of infectious motile spores (zoospores).

Water-deficient growing conditions can also importantly influence disease activity. Stress injuries, caused by desiccation and accompanying high temperatures, are often related to severe disease activity during moist periods that follow. Even though dollar spot typically requires moisture-saturated conditions, drought-stress injury before wet periods favors severe forms of it (18). Such common diseases as Fusarium blight (26, 27, 57) and Helminthosporium leafspot (15) are frequently destructive during summer on turfgrasses affected by drought. The ability of *Fusarium* spp. to grow at low-water potentials (17) and the increased sporulation by *Bipolaris sorokiniana* (Sacc. ex Sorok.) Shoem. on organic debris subjected to cyclical patterns of moisture (14, 15) have been implicated for higher levels of disease activity by these pathogens.

Most turfgrass pathogens can survive long term on dead and dying plant parts underlying the turfgrass leaf canopy. Disease is often initiated by mycelial growth or spore production by facultative fungal pathogens residing on decomposing debris. Disease suppression by saprophytic microorganisms involved in decomposition of organic debris naturally constrains development of fungal diseases (28). The ability of saprophytes to colonize rapidly the surface of decomposing plant debris, as well as to produce inhibitory substances restricting growth of other microorganisms, is strongly suppressive to most fungal pathogens (12, 15). As long as active populations of saprophytic microorganisms are intact, their presence serves to suppress continually initiation of fungal disease activities.

Moisture helps maintain populations and decomposition activities of saprophytic microorganisms on the surface of dead or dying plant debris (12). Desiccation reduces the ability of saprophytes to continue growth and often eliminates significant microorganism populations due to their sensitivity to moisture stress. In contrast, heavy applications of moisture can reduce their populations from the surface of decomposing debris by washing them into the underlying soil. Both extremes in moisture can help initiate disease activity by facultative fungal pathogens. Life cycles of a few obligate parasites, such as the rust and powdery mildew fungi, are adapted to avoid constraints of microbial suppression (13, 24). Infectious spores of obligate parasites are produced and disseminated to leaf surfaces where low populations of saprophytic microorganisms exist. Initiation of disease activity by obligate pathogens is closely related to the occurrence of favorable moisture and temperature conditions in the foliar canopy. All other fungal pathogens of turfgrasses are facultative pathogens that spend a significant part of their life cycles as saprophytes and must compete with associated microorganisms to initiate disease activity.

Influence of water conservation practices. Most common turfgrass cultivars were developed for cultural programs using abundant supplies of water. Their performance is significantly reduced in areas of restricted water use or where water penetration is impeded; they are often poorly developed and slow growing with little recuperative ability to overcome injury. Fungal disease can frequently be observed on turfgrasses grown in these areas (7). Stress caused by desiccation and high temperatures frequently increases the susceptibility of turfgrasses to common diseases. Nutrients released from desiccated turfgrass foliage can provide a food base for increased pathogen activity on drought-affected turf (42). Use of water-conserving cultural programs would restrict environmental conditions for infection to temporary periods following applications of moisture; however, timing of water applications can be important in determining the potential for disease activity. Applying water during overcast weather or at night or using frequent, light irrigation cycles often results in longer periods of leaf wetness and higher levels of fungal disease. Infrequent applications of water during the day can significantly reduce pathogen activity on turfgrass foliage.

Because of their strict requirement for lengthy periods of high moisture, most fungal diseases of turfgrasses can be reduced in incidence and severity with water-conserving cultural practices. Fusarium blight of cool-season grasses would probably be increased by water conservation practices. High temperatures and drought-affected turfgrasses are ideally suited for development of severe Fusarium blight disease (27). While reductions in water decrease the likelihood of disease attack by fungal parasites, cyclical patterns of moisture favor spore production by many facultative pathogens of turfgrasses. Spore populations of fungi causing the Helminthosporium leaf spot disease (14, 15) and Fusarium blight disease (27) are increased on turfgrasses maintained under relatively dry growing conditions. Increased applications of moisture and use of higher cutting heights are management tools for maintaining microbiological suppression of these pathogens on decomposing debris surrounding the turfgrass plant. Increased populations of spores of these pathogens should insure a higher number of infective spores on the surface of healthy leaves. Temporary periods of moisture following applications of water can result in multiple-spore infections with severe disease potential. Similar effects can be anticipated for spore deposits of such obligate parasites as the rust fungi. Applications of irrigation water during daylight hours can reduce the infection pressures of heavy spore deposits because of the rapid evaporation of water from the leaf surface.

Insects and related turfgrass pests

About 60 species of insects and mites injure turfgrasses by feeding on roots, stems, and leaves (3). Feeding can be grouped into two categories based on how insects attack the turfgrass plant: Those with chewing mouthparts devour the plant directly; sucking insects are equipped with piercing-sucking mouthparts that allow them to extract plant juices from the plant. Several sucking insects can also release toxins into plants capable of causing permanent damage to plant cells. A few insects damage turfgrasses through their nesting habits by burrowing into the soil. Other insect- and mite-related problems are more subtle and cause distorted growth, stunting, or bleaching of leaves. As a rule, extended periods of warm weather favor increasing populations and economic damage caused by insect pests. Damaging insect populations are frequently encountered during summer and fall growing seasons; however, in warmer climates, damage may continue throughout the year. Severe insect damage to lawns usually appears as patches of grass with a yellow appearance or browned areas that die if control measures are not employed.

Most turfgrass insects develop from eggs laid by adults. The morphological change that occurs from the time eggs hatch to the formation of the adult is called metamorphosis. Metamorphosis is simple for some insects and the young nymphs resemble the adults in form and feeding habits. In other insects, metamorphosis is complex; eggs hatch into larvae that do not resemble the adult form and typically feed in different habits than those of the adult. The change from the larval form to the adult is complex, involving formation of a resting stage or pupa before emergence of the adult. The larval stage of beetles (grubs), moths (caterpillars), and flies (maggots) can be very damaging to turfgrasses because of their feeding on plant parts. Frequent and thorough examination of turfgrasses is necessary to avoid severe injury caused by insect larvae. Feeding by grubs and maggots occurs below the soil line or within the thatch and is difficult to detect before development of severe injury. Foliar feeding by caterpillars frequently occurs at night and usually seems to appear overnight. Damage is usually first noticed as irregular brown spots or patches of uneven turfgrass growth.

Root-feeding insects, including white grubs, wireworms, and ground pearls, are some of the most important soil-inhabiting pests of turfgrass. With the destruction of its roots, a turfgrass is not able to obtain essential moisture and nutrients required for survival. A turfgrass injured in this manner is commonly killed when subjected to additional environmental stresses. Other soil-inhabiting organisms, such as earthworms, cicada, and cicada killer wasps, can indirectly injure turf; their tunneling and nesting often result in deposits of soil that can cover the grass.

Damage by shoot- and leaf-feeding insects varies, depending on the insect's feeding habit. Chewing insects, including sod webworms, armyworms, cutworms, billbugs, and frit flies, feed on stems and grass leaves. Removal of excessive amounts of grass tissue by heavy infestations of these insects can severely damage large areas of turf in a few days. Heavy populations of armyworms can devour established turfgrasses, creating circular bare areas. Sucking insects retard grass growth by removing plant juices from leaves and stems. Chinch bugs are considered the most damaging of sap-sucking insects attacking turfgrasses. Both nymphs and adults extract plant juices, causing grasses to turn yellow and eventually die in irregular patches. Other sap-sucking pests, such as scale insects, leafhoppers, aphids, and mites, reduce the aesthetic value or growth characteristics of turfgrasses. As with root feeders, damage caused by sap-sucking insects is more severe under environmental stress.

Water's influence on insects and related pests. While more than 60 species of insects can injure turfgrasses, only a few are sufficiently abundant to justify control.

Grubs are the most destructive soil-inhabiting insects, attacking turfgrasses in several sections of the United States (3). Grubs are the larval stage of several species of beetles. White grubs (*Phyllophaga* spp.) are the most widely distributed. They are the larval stage of June beetles that are attracted to lights in spring. In the East, grubs of Japanese beetles are also very damaging to turfgrasses. Other species that damage lawns are the Asiatic garden beetle, several species of masked chafers, European chafer, and oriental beetles. Although adult beetles of grubs differ in color, structural markings, and habits, the larvae and injuries they cause are similar.

The life cycle of *Phyllophaga* spp. varies from 2 to 4 years. Adult flight patterns are closely correlated with patterns of rainfall in spring and summer (33, 46, 47, 60). A single female can lay 50 eggs in soil. Soil moisture can influence the extent of egg deposition by the female and survival capacity of eggs. Egg deposition (20, 33, 60) and survival (20, 33) are reduced in very wet or dry soils. Following hatching, the young grubs feed on soil organic matter during the first summer and survive winter temperatures deep within the soil. Extremes in soil moisture can reduce their survival in soil (21, 33). During successive growing seasons, grubs actively feed on grass roots until they enter the pupae stage. Moist soils are thought to be more conducive to tunnelling and feeding than are drier soils. In severe infestations, turfgrass sod can be lifted or rolled up like a carpet. Moisture stress usually kills large areas of turfgrass following extensive feeding by grubs. Frequent applications of water are needed to keep the turfgrass alive until an insecticide can be employed to control the pests and to encourage development of new roots.

Mole crickets are increasingly important pests of turfgrasses in the southeastern United States and have been observed as far west as Texas (50, 55). By burrowing they uproot turfgrasses in sandy soil and the plants wither and die. Mole crickets become inactive in dry soil and typically cease surface burrowing until soil moisture conditions are more favorable. The nest-building habits of several species of ants and a few bees and wasps are also troublesome to lawns. Burrowing and soil mounding of these insects are also increased when soils are moist.

Caterpillars of several insect species frequently damage turfgrasses. These are larvae of moths. Sod webworms and armyworms are considered the most important of the caterpillars attacking turfgrasses. There are many species of sod webworms widely distributed in the United States (3). Sod webworms can have several generations per year, depending on the species and location. Larvae feed on turfgrass leaves and shoots at night and during cloudy weather. The fall armyworm feeds on leaves and can devour the entire foliar canopy. This insect produces several generations per year in the southern part of its range. Larvae of both of these insects burrow into the soil during inactive periods of the day to escape high temperatures and desiccation injury. Survival of larvae is increased by frequent applications of moisture, which could partially explain why the pests are more damaging on golf greens during summer months.

Several **chinch bug** species are important sap-sucking pests of turfgrasses (48, 51). Both nymphs and adults suck plant juices and cause the grass to turn yellow in irregular patches. Eventually turfgrasses affected by high populations will turn brown and die. Several species of cool-season grasses and St. Augustinegrass grown in the South are damaged by chinch bugs. Feeding activities occur during hot summer months when the turf is subjected to moisture stress. *Beaveria bassiana*, a biological agent that feeds on chinch bugs, is favored by moist environments (49). Decreased moisture during summer favors the buildup of populations when natural enemies are at low levels.

Feeding by **scale insects** is increased when low levels of moisture are supplied to turfgrasses. Some scale insect species attack the roots; others damage leaves and stems. Newly hatched scales, called crawlers, actively move about over the turfgrass plant. Within a few hours, they establish a permanent feeding site, lose their legs, and remain attached to the plant. Their increased activities in dry environments can be related to increased damage from the removal of plant juices during water stress or a reduction of predatory insects in dry climates. The bermudagrass stunt mite is also a common sap-sucking insect during hot summers. Reproduction and feeding by the stunt mite are favored by dry climates and high temperatures.

Influence of water conservation practices. Feeding by several species of insect pests can be reduced in turfgrasses grown under water-conserving cultural programs. The mobility and survival of soil-inhabiting insects can be greatly reduced in dry environments. Lengthy irrigation cycles that promote growth of deep turfgrass root systems probably encourage root feeding by soil insects in deeper soil profiles less subject to drying. Egg laying, larval survival, and feeding by root-feeding grubs are adversely affected by dry soils. Water conservation practices can significantly reduce populations and activities by these pests. Burrowing and nesting of soil insects, such as ants, earthworms, mole crickets, and cicadas, can also be restricted in dry soil environments; however, these pests are well adapted to take advantage of temporary periods of moisture to continue soil mounding or burrowing.

A continuous exchange of moisture exists between insects and the humid atmosphere (37). Some insect species can tolerate dry environments for extended periods; others are more sensitive to desiccation injury. Dry growing conditions are generally favorable for the development of such sap-sucking insects as chinch bugs, scale insects, leaf hoppers, and eriophyid mites. Increased populations and feeding by chinch bugs are frequently favored by dry climates. Damage caused by the feeding of nymphs and adult chinch bugs is more severe where turfgrasses receive limited moisture. Heavy population pressures of chinch bugs during summer potentially affect several cool-season grasses and St. Augustinegrass grown in the South. Stunted growth of bermudagrass caused by the feeding of the bermudagrass stunt mite is also common during summer's drier periods. High temperatures and dry climates favor reproduction and feeding by the pest.

Nematode diseases

While most soil-inhabiting nematodes are considered beneficial because they feed on soil microorganisms and decaying organic matter, some are important parasites of higher plants, including turfgrasses (41). Plant parasitic nematodes are microscopic worms that feed only on living plant tissue and are rarely found on dead roots. Parasitic forms are equipped with a specialized feeding structure, the stylet, which is used for puncturing plant cells and withdrawing liquified cell contents. Enzyme secretions by some nematode species cause abnormalities in cell growth, which reduce the functional ability of roots or other plant organs. Large populations of plant parasitic nematodes feeding on roots can cause turfgrass plants to decline slowly. Injuries by these pests can occur on all types of turfgrasses; however, they seldom kill them. The interactive influence of nematode injury and soil-borne fungal pathogens can devastate turfgrasses and is considered an important aspect of their destructiveness (64). Severe nematode infestations are commonly associated with stunted growth, yellowing of leaves, and reduced turfgrass vigor during summer. Nematode damage is usually restricted to sandy soil with excessive moisture. Nematodes have a greater mobility in porous soils where there is a thin film of water around soil particles. Heavy clay soils or soils maintained in a relatively dry state restrict nematode movement and feeding.

Based on their feeding habits and association with the host plant, parasitic nematodes are either endoparasitic or ectoparasitic. Endoparasitic forms invade roots and spend all or part of their life cycles within plant tissue; the cyst, cystoid, root knot, lesion, and burrowing nematodes are examples of this group. Ectoparasitic types or free-living nematodes feed on the surface of plant roots and spend their life cycles in the soil. Most of the important nematodes attacking turfgrasses are ectoparasitic (64). The **spiral**, **sting**, **stunt**, **ring**, **pin**, **stubby root**, **dagger**, **lance**, and **needle nematodes** are examples of this group.

All nematodes reproduce sexually. The nematode life cycle can be separated into six stages: the egg, four larval stages, and the adult (10). Nematode larvae hatch from eggs deposited in the soil or in surface tissue of roots. As the larvae mature, each larval stage is terminated by a molting of the cuticle, similar to that occurring in insect larvae. The life cycle usually lasts for 30 to 45 days (63, 64), depending on temperature, moisture, suitability of the host, and the individual species.

No resistant turfgrass cultivars are available with resistance to all parasitic nematodes (5). Applications of nematicides help reduce high populations of parasitic nematodes within the root zone of affected turfgrasses. The effect of nematicides is temporary and often requires repeated applications where soil conditions favor the reentry and buildup of nematode populations from underlying soil.

Water influences nematode diseases. Soilborne nematodes are water loving with a high dependency on free water in soils for their mobility, parasitic activities, and reproductive ability (10). Moist, sandy soils of porous texture favor root-feeding species; heavier soils restrict their movement and feeding. The cuticle or outer covering of nematodes is permeable to water; consequently, their interior water balance in dry environments cannot be maintained (41). In moist soils these parasites are damaging to turfgrasses; in the absence of free water their role as problem pests is greatly reduced.

Methods for detecting parasitic nematodes in soil rely on laboratory extraction techniques to determine populations of specific nematode species, particularly in the upper soil profile. Although pathogenicity of various nematode species has been demonstrated repeatedly, consistent and clearly defined correlations of nematode populations with turfgrass damage are difficult to obtain (64). Water-conservation practices for turfgrasses can restrict nematode populations and feeding to the deeper soil profiles. Only a few studies have investigated nematode populations in turfgrass soils at depths greater than 15 cm. Ectoparasitic forms including pin (*Paratylencus* spp.), stubby root (*Trichodorus* spp.), and dagger (*Xiphinema* spp.) nematodes and the endoparasite lesion (*Pratylenchus* spp.) nematode have been recovered at depths exceeding 25 cm; however, the extent of damage caused by root feeding at these depths has not been clearly defined.

Pest control strategy with limited water resources

Turfgrass pest control programs are focused on managing pest populations below objectionable damage levels for cultivars in use. Few turfgrass pests can be eliminated long from the turfgrass community. The invasive characteristics of many pests are sufficient to provide reentry pressures when environmental conditions or cultural programs favor their development. The continuing presence of other pests with limited mobility is aided by their dispersal on turfgrass cultural equipment, clippings, soil, foot traffic, and movement by wind and water. Turfgrass pests are also well equipped with mechanisms for insuring their survival. Resting structures, such as weed seeds, dormant spores and sclerotia of fungal pathogens, and eggs or larvae of insects and nematodes, are capable of long-term survival in harsh environments. Thus, turfgrass pest populations can be viewed as a continuing threat throughout the year, with regulatory influences of environment, turfgrass species, and cultural variables their primary constraints.

Harsh environments created by turfgrass water conservation practices can significantly reduce numbers of important pests attacking turfgrasses. Injuries caused by a few pests, however, can damage turfgrass plants weakened by these same environments. Higher temperatures and reduced humidities within the turfgrass leaf canopy are consequences of restricted water use that can reduce the growth potential of common turfgrass cultivars. As a rule, cool-season turfgrass species are poorly adapted to withstand stresses caused by high temperature and desiccation. Growth of stoloniferous warm-season grasses can also be significantly reduced by limited availability of water. Pests adapted to relatively dry environments, with temporary periods of available moisture, can become very damaging when water is limited. Pests attacking weak-growing turf during droughts demonstrate this effect. Nigrospora Stolon dieback of St. Augustinegrass (16) was observed in Texas during a long drought in association with fungal lesions on stolon internodes. The girdling activity of the lesions limited water supplies to terminal ends of unrooted stolons and caused severe dieback. Where water use is not restricted, the disease is seldom observed.

With restricted water use, the duration and timing of irrigation cycles are important considerations for limiting turfgrass pest pressures. Frequent, light irrigation apparently aids development of most turfgrass pests; the resulting surface moisture encourages weed seed germination and seedling survival. Frequent irrigation also insures the presence of humid environments favorable for feeding and survival of most insect pests. Cyclical patterns of wetting and drying resulting from frequent, light watering can enhance fungal spore production and rapid infection by spores deposited on drought-stressed plant parts. Fungal pathogens capable of producing many infective spores are well equipped to take advantage of temporary periods of moisture. Spores produced by *D. sorokiniana* are capable of germinating in 1 hour in the presence of free water (25); spores of fungi causing the rust diseases require 4 to 8 hours of moisture for infection (18). Timing of irrigation can also significantly influence moisture retention within the turfgrass foliar canopy. Irrigation in the morning frequently reduces retention of moisture in the leaf canopy because evaporation occurs at higher rates during the day. In contrast, irrigation at night favors long-term moisture retention in the leaf canopy. The presence of long moist periods and moderate temperatures following irrigation at night could significantly increase populations and activities of turfgrass pests.

Use of reduced levels of fertility on turfgrasses maintained with limited supplies of water could contribute to reduced populations and activities of many turfgrass pests. Nitrogen, a limiting factor in the decomposition of organic debris, is also a limiting factor for saprophytic growth of facultative fungal pathogens. Saprophytic development and spore production by fungal pathogens on decomposing debris can be reduced by limiting supplies of nitrogen available for utilization by these fungi (14). Reduced levels of fertility can also decrease production of succulent plant growth susceptible to damage by many disease and insect pests. Problem weeds compete with turfgrasses for available supplies of nutrients in the soil. Decreased levels of fertility commonly reduce the rate of establishment of tender weed seedlings and encourage their susceptibility to destruction during periods of environmental extremes.

Higher mowing height can be beneficial in reducing populations and injuries of pests on turfgrasses maintained with limited water. Although transpirational water loss is increased by higher cutting heights (45, 54), the shading effect of extended leaves can reduce temperature and moisture stress caused by water-deficient growing conditions. This factor could be an important determinant for sustained growth of many turfgrass cultivars. Shading by extended leaves encourages water conservation and moderate temperatures on surface debris, favoring decomposition and disease-suppressive activities of saprophytic microorganisms. Shading can also reduce seed germination and establishment of weed seedlings. Insect pests may have a higher capacity for survival under the leaf canopy of higher cut turfgrasses; however, moderate environments created by higher cutting heights could increase the numbers of natural predators and other biological agents capable of attacking these pests.

The need for pesticides on water-conserving turfgrass cultural systems could probably be reduced significantly. Reductions in turfgrass pest populations and their activities can reduce demand for pesticide applications to control relatively few pests. While reductions in total demand can be anticipated, higher demands for target-specific pesticides may occur to control specific pests adapted to dry environments. Application of most turfgrass pesticides relies on recovery of actively growing plants to achieve desired results. The slow recuperative ability of turfgrasses receiving limited water can restrict the benefits of many pesticides. High temperatures and low humidities do not favor uptake of pesticides by the turfgrass plant or the target pests. These environmental factors reduce the effective period for pesticide activity by increasing the rate of volatilization and photodecomposition of surface-applied chemicals. Application of pesticides to dry-leaf surfaces may increase the concentration requirements for effective pest control or increase the volume of spray necessary to insure their adequate distribution. Toxicity of pesticides to turfgrasses could also be increased by application in hot, dry environments. Slow-growing plants subjected to environmental stress have little ability to tolerate additional stress. Applications of pesticides during evening or early morning could prove a satisfactory alternative for insuring effectiveness of treatments.

Integrated pest management (IPM) programs for turfgrasses are based on the use of sound turfgrass management programs to limit populations of pests using cultural, biological, and chemical control strategies (56). One principal benefit of IPM is the use of alternative pest control measures where pesticides are used extensively. Although the IPM concept is relatively new to turfgrass pest management strategy, many IPM principles have been recommended (22) and effectively employed by turfgrass managers for decades. Development of vigorously growing and healthy turfgrass stands is fundamental to the success of turfgrass IPM strategies. Cultural variables and biological pest management capabilities of IPM strategy could be significantly restricted under water conservation programs for turfgrasses. Use of vigorous turfgrass growth as a pest management strategy would be severely affected by reductions in turfgrass growth capabilities under water-restrictive growing conditions. Long-term effects of water conservation would call for development of turfgrass cultivars with tolerance or resistance to specific pests. These cultivars could be widely used for environmentally safe methods of pest control.

Literature cited

1. ALKERE, W. K. II
    1982. Problem soils for problem waters. Proc. The 1982 Turf and Landscape Institute. Univ. of Calif.
2. ANON.
    1977. Weed infestations brought about by overwatering. Broward County Agric. Agents Office, Fla.
3. APP, B. A., and S. H. KERR
    1969. Harmful insects. In A. A. Hanson and F. V. Juska (eds.), Turfgrass Science, Monograph 14. Amer. Soc. of Agron., Madison, Wis. 715 pp.
4. BAYER, D. E.
    1967. Effect of surfactants on leaching of substituted urea herbicides in soil. Weed Sci. 15:249–252.
5. BEARD, J. B.
    1973. Turfgrass Science and Culture. Prentice Hall, Englewood Cliffs, N.J. 658 pp.
6. BEARD, J. B., P. G. RIEKE, A. J. TURGEON, and J. M. VARGAS
    1978. Annual bluegrass (*Poa annua* L.). Descriptions, adaptive culture and control. Res. Rpt. 352. Mich. State Univ. Agric. Expt. Sta. 31.
7. BENNETT, O. L., and E. S. ELLIOTT
    1972. Plant disease incidence on five forage species as affected by north- and south-facing slopes. Plant Dis. Rptr. 56:371–375.
8. BURNS, WENDELL
    1982. Fertilizers, chemicals, soil conditioners. Retailers' guide to selling turf products. Lawn and Garden Marketing. February. 14–15.
9. BURPEE, L. L.
    1980. Identification of *Rhizoctonia* spp. associated with turfgrasses. In B. G. Joyner and P. O. Larsen (eds.), Advances in Turfgrass Pathology. Harcourt Brace Jovanovich, Inc., Duluth, Minn. 197 pp.
10. CHRISTIE, J. R.
    1959. Plant nematodes—their bionomics and control. Univ. Fla. Agri. Expt. Sta., Gainesville, Fla.
11. COCHRANE, V. E.
    1958. Physiology of Fungi. Wiley and Co., N.Y. 524 pp.
12. COLBAUGH, P. F.
    1973. Environmental and microbiological factors associated with saprophytic and parasitic activities of *Helminthosporium sativum* on *Poa pratensis*. Ph.D. thesis. Univ. of Calif., Riverside, 93 pp.
13. ———
    1980. Cultural practices to control fungus diseases in turf. Grounds Maintenance. Vol. 15(1), pp. 48–54.
14. COLBAUGH, P. F., and J. B. BEARD
    1980. Influence of cultural practices on conidial production by *Drechslera* and *Bipolaris* spp. In B. J. Joyner and P. O. Larsen (eds.), Advances in Turfgrass Pathology. Harcourt Brace Jovanovich, Inc., Duluth, Minn. 197 pp.

15. COLBAUGH, P. F. and R. M. ENDO
    1974. Drought stress: An important factor stimulating the development of *Helminthosporium sativum* on Kentucky bluegrass. Proc. Second Intl. Turfgrass Res. Conf. 328–334.
16. COLBAUGH, P. F., and S. J. TERRELL
    1982. Nigrospora stolon rot of St. Augustinegrass related to drought conditions in Texas. Grounds Maintenance. April. 40–41.
17. COOK, R. J., and R. I. PAPENDICK
    1970. Soil water potential as a factor in the ecology of *Fusarium roseum* F. sp. cerealis "Culmorum." Plant and Soil. 32:131–145.
18. COUCH, H. B.
    1974. Diseases of Turfgrasses. 2nd ed. R. E. Krieger Publ. Co., Huntington, N.Y. 348 pp.
19. DAVIDSON, R. H., and M. P. LEONARD
    1967. Insect Pests of Farm, Garden, and Orchard. John Wiley and Sons, N.Y. 675 pp.
20. DAVIDSON, R. L., and R. J. ROBERTS
    1968. Influence of plants, manure, and soil moisture on survival and liveweight gain of two scarbaeid larvae. Entomol. Exp. Appl. 11:305–314.
21. DAVIDSON, R. L., J. R. WISEMAN, and V. J. WOLFE
    1972. Environmental stress in the pasture scarab *Sericesthis nigrolineata* II. Effects of soil moisture and temperature on survival of first-instar larvae. J. Appl. Ecol. 9:799–806.
22. ELMORE, C. L., and W. B. McHENRY
    1980. *In* Turfgrass Pests. Priced Publication No. 4053. Univ. of Calif. 53 pp.
23. ELMORE, C. L.
    1983. Working paper on "Urban Pest Management" for Urban Pest Management Nat. Academy Press, Washington, D.C.
24. ENDO, R. M.
    1972. The turfgrass community as an environment for the development of facultative fungal parasites. *In* V. B. Youngner (ed.), The Biology and Utilization of Grasses. Academic Press, N.Y. 426 pp.
25. ENDO, R. M., and R. H. AMACHER
    1964. Influence of guttation fluid on infection structures of *Helminthosporium sorokinianum*. Phytopathology 54:1327–1334.
26. ENDO, R. M., R. BALDWIN, S. COCKERHAM, P. F. COLBAUGH, A. H. McCAIN, and V. A. GIBEAULT
    1973. Fusarium blight, a destructive disease of Kentucky bluegrass and its control. Calif. Turfgrass Culture 23:1–2.
27. ENDO, R. M., and P. F. COLBAUGH
    1974. Fusarium blight of Kentucky bluegrass in California. Proc. 2nd Intl. Turfgrass Res. Conf. 325–327.
28. ———
    1974. Biological control of turfgrass disease—an unrealized potential. Golf Superintendent. June. 21–24.
29. ENDO, R. M., and J. J. OERTILI
    1964. Stimulation of fungal infection of bentgrass. Nature. 201:313.
30. ENGEL, R. E., and R. D. ILNICKI
    1969. Turf weeds and their control. Chap. 9. *In* A. A. Hanson and F. V. Juska (eds.), Turfgrass Science. Monograph 14. Amer. Soc. of Agron., Madison, Wis. 715 pp.
31. FREEMAN, T. E.
    1980. Seedling diseases of turfgrasses incited by *Pythium* spp. *In* B. G. Joyner and P. O. Larsen (eds.), Advances in Turfgrass Pathology. Harcourt Brace Jovanovich, Inc., Duluth, Minn. 197 pp.
32. FREEMAN, T. E., and G. C. HORN
    1963. Reaction of turfgrasses to attack by *Pythium aphanidermatum*. Plant Dis. Rptr. 47:425–427.
33. GAYLOR, M. J., and G. W. FRANKIE
    1979. The relationship of rainfall to adult flight activity and of soil moisture to oviposition behavior and egg and first instar survival in *Phyllophaga crinita*. Environ. Entomol. 8:591–594.
34. GIBEAULT, V. A.
    1970. Perenniality in *Poa annua* L. Ph.D. Dissertation. Oreg. State Univ. 124 pp.
35. GIBEAULT, V. A., R. AUTIO, S. SPAULDING, and V. B. YOUNGNER
    1982. Seed mixes for Fusarium blight. Plant Dis. Rptr. 66:265.
36. GIBEAULT, V. A.
    1983. Personal communication. Botany and Plant Sciences Department, Univ. of Calif., Riverside.
37. GOVAERTS, J., and J. LECLERCQ
    1946. Water exchange between insects and air moisture. Nature. London. 157:483.
38. GUTTADUARIO, J., E. E. HARDY, and A. S. LIEBERMAN
    1978. An investigation of turfgrass land use acreages and selected maintenance expenditures across New York State. Cornell Univ. Dept. Floriculture and Orn. Hort. Rpt. 36 pp.
39. HOLT, E. C.
    1969. Turfgrasses under warm humid conditions. Chap. 20. *In* A. A. Hanson and F. V. Juska (eds.), Turfgrass Science. Monograph 14. Amer. Soc. of Agron., Madison, Wis. 715 pp.
40. HOLT, J. S.
    1983. Personal communication. Botany and Plant Sciences Department, Univ. of Calif., Riverside.
41. JONES, F. G. W.
    1959. Ecological relationships of nematodes. *In* C. S. Holton (ed.), Plant Pathology, Problems and Progress 1908–1958. Univ. of Wis. Press, Madison, Wis. 588 pp.
42. KEMBLE, A. R., and H. T. MacPHERSON
    1954. Liberation of amino acids in perennial ryegrass during wilting. Biochem. 58:46–49.
43. LANGE, A. H.
    1983. Personal communication. Extension weed scientist, Univ. of Calif., Kearney Agric. Center, Parlier, Calif.
44. MADISON, J. H.
    1962. The effect of management practices on invasion of lawn turf by bermudagrass [*Cynodon dactylon* (L.)]. Amer. Soc. Hort. Sci. 80:559–564.
45. MADISON, J. H., and R. M. HAGAN
    1962. Extraction of soil moisture by Merion bluegrass (*Poa pratensis* L. 'Merion') turf, as affected by irrigation frequency, mowing height, and other cultural operations. Agron. J. 54:157–160.
46. POTTER, D. A.
    1981. Seasonal emergence and flight of northern and southern masked chafers in relation to air and soil temperature and rainfall patterns. Environ. Entomol. 10:793–797.
47. ———
    1983. Effect of soil moisture on oviposition, water absorption and survival of southern masked chafer (Coleoptera: Scharbaeidae) eggs. Environ. Entomol. 12:1223–1227.
48. RATCLIFFE, R. H.
    1982. Evaluation of cool-season turfgrasses for resistance to the hairy chinch bug. *In* H. D. Niemczyk and B. G. Joyner (eds.), Advances in Turfgrass Entomology. Hammer Graphics, Piqua, Ohio. 150 pp.
49. REINERT, J. A.
    1978. Natural enemy complex of the southern chinch bug in Florida. Ann. Entomol. Soc. America. 71:728–731.
50. ———
    1980. Southern turfgrass insect pests with emphasis on mole cricket biology and management. Proc. Fla. Turfgrass Management Conf. 28:33–42.
51. ———
    1982. A review of host resistance in turfgrasses to insects and Acaradines with emphasis on the southern chinch bug. *In* H. D. Niemczyk and B. G. Joyner (eds.), Advances in Turfgrass Entomology. Hammer Graphics, Inc., Piqua, Ohio. 150 pp.
52. ROBERTS, E. C., F. E. MARKLAND, and H. M. PELLET
    1966. Effects of bluegrass stand and watering regime on control of crabgrass with preemergence herbicides. Weed Sci. 14:157–161.
53. SALADINI, J. L.
    1980. Cool- versus warm-season Pythium blight and other related Pythium problems. *In* B. G. Joyner and P. O. Larsen (eds.), Advances in Turfgrass Pathology. Harcourt Brace Jovanovich, Inc., Duluth, Minn. 197 pp.
54. SHEARMAN, R. C., and J. B. BEARD
    1973. Environmental and cultural pre-conditioning effects on the water-use rate of *Agrostis palustris* Huds., cultivar Penncross. Crop Sci. 13:424–427.
55. SHORT, D. E., and J. A. REINERT
    1982. Biology and control of mole crickets in Florida. *In* H. D. Niemczyk and B. G. Joyner (eds.), Advances in Turfgrass Entomol-

ogy. Hammer Graphics, Inc., Piqua, Ohio. 150 pp.
56. SHORT, D. E., J. A. REINERT, and R. A. ATILANO
    1982. Integrated pest management for urban turfgrass culture in Florida. *In* H. D. Niemczyk and B. G. Joyner (eds.), Advances in Turfgrass Entomology. Hammer Graphics, Inc., Piqua, Ohio. 150 pp.
57. SMILEY, R. W.
    1980. Fusarium blight of Kentucky bluegrass, new perspectives. *In* B. G. Joyner and P. O. Larsen (eds.), Advances in Turfgrass Pathology. Harcourt Brace Jovanovich, Inc., Duluth, Minn. 197 pp.
58. ———
    1983. Compendium of Turfgrass Diseases. Amer. Phytopathol. Soc., St. Paul, Minn. 102 pp.
59. SPRAGUE, H. B., and E. E. EVAUL
    1930. Experiments with turfgrasses in New Jersey. N.J. Agr. Exp. Sta. Bull. No. 497. 55 pp.
60. SWEETMAN, H. L.
    1931. Preliminary report on the physical ecology of certain *Phyllophaga*. Ecology 12:401–422.
61. TOOLE, E. H., and V. K. TOOLE
    1941. Progress of germination of seed of *Digitaria* as influenced by germination temperature and other factors. J. Agr. Res. 63:65–90.
62. TURGEON, A. J.
    1980. Turfgrass Management. Reston Publ. Co., Reston, Va. 391 pp.
63. TURGEON, A. J.
    1982. Turfgrass pest management. *In* R. W. Sheard (ed.), Proc. Intl. Turfgrass Res. Conf. Guelph, Ontario, Canada. 351–368.
64. VARGAS, J. M., JR.
    1981. Management of Turfgrass Diseases. Burgess Co., Minneapolis, Minn. 204 pp.
65. WATSON, J. R.
    1972. Effects on turfgrasses of cultural practices in relation to microclimate. Chap. 14. *In* V. B. Youngner (ed.), The Biology and Utilization of Grasses. Academic Press, N.Y. 426 pp.
66. WELTON, F. A., J. C. CARROLL, and J. D. WILSON
    1934. Artificial watering of lawn grass. Ecology 15:380–387.
67. WILSON, J. D.
    1927. The measurement and interpretation of the water supplying power of the soil with special reference to lawn grasses and some other plants. Plant Phys. 2:385–440.
68. WILSON, C. G., and J. M. LATHAM, JR.
    1969. Golf fairways, tees, and roughs. *In* A. A. Hanson and F. V. Juska (eds.), Turfgrass Science. Monograph 14. Amer. Soc. of Agron., Madison, Wis. 715 pp.
69. YOUNGNER, V. B.
    1959. Ecological studies on *Poa annua* in turfgrasses. J. Brit. Grassland Soc. 14(4):233–247.
70. YOUNGNER, V. B., A. W. MARSH, R. A. STROHMAN, V. A. GIBEAULT, and S. SPAULDING
    1981. Water use and turfgrasses. Calif. Turfgrass Culture 31:1–4.

Practicum

11 Site design for water conservation

CAL O. OLSON

Cal O. Olson is the founder of Olson Associates, Newport Beach, California. He is a golf course architect registered as a professional engineer and landscape architect. His 23 years of experience in irrigation and landscape water use includes industrial and commercial areas, public facilities, new communities, recreational developments, golf courses, and resorts.

Water conservation and water management require the ability to measure results and compare these with established goals. Site design that implements standards required to manage a system is essential to achieving this goal. Reduced water use in itself does not equate to water conservation.

Irrigation schedules and programs are related directly to evapotranspiration rates for the area and the plant material being irrigated.

Irrigation requirements increase as evapotranspiration rates increase. Application rates for the irrigation systems should be uniform within any given control system so time of operation can be calculated to supply supplemental water as required.

System design should allow separation of control valves based upon soil types, landscape material, and terrain variances to allow further adjustment of time schedules to accommodate these variables.

Irrigation designers interested in providing systems that permit water conservation and water management policies should provide the managers with an appropriate system. Such a system gives the irrigation manager the ability to monitor, control, measure, and establish feedback information to help him make decisions effectively.

11 Site design for water conservation

Cal O. Olson

This section offers practical techniques and recommendations in developing a well-managed irrigation system. Often, a consultant is called to help develop a management program after the irrigation systems are in the ground. While this can be successful in some cases, it is obviously better to plan the system from the beginning, utilizing water management goals and established irrigation principles. The key to the properly managed system lies within the five categories presented here. Failure to implement or understand any portion can result in a less than perfect system. This approach to design does not guarantee less water use, correct water use, or resource planning—only the tools to provide these possibilities. The categories:

- Establishing design criteria
- Irrigation master planning
- Implementation
- Maintenance guidelines
- Water management programs

Establishing design criteria

Design criteria, the foundation for a well-managed and well-understood system, should be well thought out. The list of considerations is lengthy, and each consideration is dealt with briefly, focusing on the most important aspects.

Water availability. Sufficient quantity and quality of water available for a project are obviously required to meet plant material requirements and soil interactions. Water quality should be analyzed, especially when it comes from a nondomestic or reclaimed water source (i.e., filters, effects on soil and plants, equipment selection, etc.). Preparation of practical water quantity requires a master plan. The following graph depicts the relationship between water availability and demand.

Maintenance. Irrigation design criteria and the final design should incorporate maintenance input. For example, particular mowing schedules that could affect timing of sprinkler operations or a golf course's maintenance could include fertilizing programs and syringe programs that could affect head placement, control programming, valve sectioning and placement, and other important aspects of sprinkler design or hydraulics. For example, park maintenance differs from golf maintenance or from linear median-strip street maintenance. The designer should investigate the maintenance program thoroughly and incorporate it into the design. If no criteria are available, the designer should establish a capability program and submit the program to the owner, including the capabilities required of the manager.

Codes and regulations. Governing codes and regulations vary throughout the United States, and such items as backflow prevention, reclaimed water use, pumping plants, metering, etc., usually are regulated. These should be incorporated into contract documents and design criteria. Agency submittals and required formats should also be investigated for proper compliance.

Soils. The designer must evaluate the project's soils. A large project can have varying types of soils that require several design decisions within one site. Different soil types can affect operation schedules, system sectioning, nozzle selection, etc. The site should also be analyzed for cut-and-fill areas, as percolation rates generally vary within these areas, even when soil types are similar.

Engineering coordination. The irrigation designer must coordinate utilities, such as water and power and sleeving, with the project engineer in a timely manner. These requirements can be

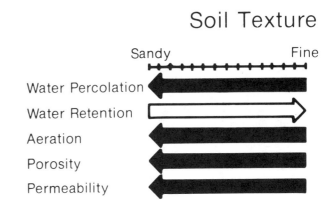

Soil considerations and qualities. Direction of arrow indicates increase.

coordinated after contract documents are completed (if timely), or they may be evaluated at the irrigation master plan level for various alternatives. Recommended timing for utility studies and plans is during the irrigation master plan stage.

System controls. The control system ultimately selected for the project should be evaluated, based on the requirements of the particular project. Controls and valves are the heart of the system and should be carefully evaluated. One can evaluate, for example, through a matrix of types of controls such as manual, conventional time clock, satellite controls, moisture-sensing microprocessor network, computer control, or others. On the other axis of the matrix, the particular governing factors can be numbered and weighted according to priority. Smaller projects are relatively straightforward, and valve and control selection does not require a major decision. Projects bigger and more complex require controls programmed for efficiency and to account for potentially complex hydraulic maintenance and germination considerations. Typical considerations (not in any priority order) are:

- Energy efficiency
- Capitalization costs
- Operating costs
- Longevity
- Operator skill level required
- Simplicity of maintenance or required maintenance
- Replacement part availability
- "State-of-the-art" concepts
- Ability to phase in large projects
- Power-surge protection
- Fault-signal relay
- Wiring simplicity
- Local contractor familiarity
- Local manufacturer representation
- Large areas effectively irrigated
- Central-control capabilities
- Video readout
- Quick reaction to climatic changes
- "Down-time" effects
- Owner's desires

Controller functions and programming possibilities. First, one should discuss the various controls and their functions. Electromechanical controls date back many years and are not new to irrigation. A typical control unit simply activates the valves sequentially for the duration setting for each station. There are many variations of control units, from simple to sophisticated. The more simple control units start once a day at a preset time (usually at a pin for the hour and at a pin for the day), and each valve will function for the preset time. Individual preset times have maximum durations of 30 to 60 minutes. Other capabilities repeat times, omit station possibilities, use a calendar wheel, etc.

General classification of controls. A general classification of controls would be programmers, sensor-based programmers, modular or satellite systems, and computer control systems. Programmers (usually called controllers or clocks) are "stand-alone" units that sequentially activate valves on scheduled programs. Sensor-based programmers either allow their programs to be overridden upon receiving signals from moisture sensors that the indicator point requires no water or they have no real time setting except a start time and take their command from sensors to activate the valves sequentially. Modular or satellite systems have a central (or master) controller that activates the field (or slave) units. The field unit then activates the valves wired to the unit sequentially. Computer-controlled systems vary from visual CRT units, with unlimited flexibility, to central "line" readouts that control field units that operate valves sequentially.

Programmers. Controls having repeat capabilities vary by manufacturer for the flexibility desired. Some controls repeat at the end of the original cycle back to the beginning immediately, and water all those stations designated with "repeat." Some controls can operate more than once a day, but all stations must operate; others can select those that operate on the repeat and those that do not.

Some controls offer dual programming for turf and shrubs, and the more sophisticated controls have more flexibility in multiprogramming capabilities. Some control units offer three different control unit functions within one; each of the three programs can operate on independent days and each station can function from any of the three programs with a dial indicator. In addition, variations to the controller can be specified to include moisture-sensing capabilities for each cycle. This particular variation, developed for slope irrigation, functions well in other areas to account for exposures and conserving on water consumption. Another variation is the "sentry system" that allows maintenance personnel to review the system's previous operation. If one station does not operate, the controller indicates this malfunction with indicator lights at the control unit.

Electromechanical control units provide a day-and-time switch for each control station, providing flexibility in operating each station as if it were on a separate control unit.

Satellite systems. The modular-satellite system has been used generally for large parks, housing developments, or golf courses. In this instance, the master control activates the satellite controls, usually through low-voltage wires. Satellite controls are provided with a duration timing for each control station only and do not have a day or hour wheel. The day-and-hour wheels are at the master unit. This type of system is especially useful when a central control is desired or the designer is ensuring that systems will control the hydraulics of a large distribution system.

Long-duration systems. Control units for long-duration operation are generally used for drip systems and are usually of the agricultural, computer, or microprocessor type. Some control units have added multiplier systems to their controls to provide up to eight times the duration shown on the station-control timing.

Digital solid state. Numerous controls have digital display and timing. Their functions are offered in the electro-mechanical-type controls, but timing is more precise, and less costly. Digital and solid-state controls should be protected from the elements.

Computer/microprocessor controls. Computer and microprocessor control units have proved effective in today's market, especially because of the water-conservation techniques available through their use. The sophistication of each of these types varies and proper selection would best be considered through a matrix decision. A few of the options available now include microprocessor readouts back to a central monitoring location, with moisture sensing throughout; "line"- reading digital controls controlling field units, and desktop CRT units programmed in groups with varying degrees of feedback on printers or data input. Large projects should evaluate their use and efficiency carefully. No one should arbitrarily select any for a project without honestly evaluating project needs, options, and plans for the future.

Considerations. Some key factors to consider: Does the unit have the flexibility for your various plant palette selections, types of material of ground cover, and frequency of watering required? Do slope conditions require short-duration and multiple-repeat applications, repeat selection at preset times, and repeat selection for some stations and not for others on the same controller? And is there proper station duration timing as your design requires? Maintenance input meeting both operational and germination requirements is key to ultimate program requirements.

Summary of types. According to their construction, controls may be classified as mechanical; electromechanical; solid-state, digital controllers; or microprocessor computer-controlled. Mechanical controls have motors that turn cams, which in turn actuate levers (microswitches); this type of control is generally used in residential application because of its low manufacturing cost and the low flexibility requirements of residential owners. Electromechanical controls have motors that turn rotary switches (wafers) to actuate relays and solid-state circuits; this type is often specified for commercial and medium-scale developments because of its simplicity of operation, reasonable dependability, and adequate flexibility. Solid-state, digital controllers are available in conventional programming (digital readout) or conventional programming with modifications, keyboard microprocessor-based or hybrids with card reader types or modified minicomputers; these are generally used for medium-scale projects and when maintenance personnel are concerned with water management at a higher level. The computer controls offered are generally used for parks, golf courses, and large-scale developments with a sophisticated level of maintenance personnel, management, and owner concern.

Once a control system has been selected, a matrix can be developed evaluating the various manufactured products available on one axis, and the other axis can be established using desired capabilities of the system and evaluated in a similar manner as evaluating the type of control system desired. If a computer control system is desired, possible considerations could include:

- Printer availability
- Video readout
- Synoptic field-map capability
- Quick reaction to climate capability
- Ability to expand for future
- Warranty and service
- Operation simplicity
- Programming simplicity
- Program flexibility
- Phasing capabilities
- Sequential-control operation only
- Complete program control options
- Ability to change timing by percentage function
- Quality of compatible control valve
- Field-operation potentials
- "Overriding" input potential (i.e., weather)
- Capitalization costs (cost/benefit ratio)
- Surge protection
- "State-of-the-art" concepts
- Energy efficiency
- Owner's desires
- Miscellaneous (designs to evaluate)

There are other methods for selecting a control system, but this checklist has proved effective. Considerations can differ for each project, and the items selected should be carefully reviewed. In any case, design criteria should be justified, knowing all the parameters.

Valves. References to the controlling of a "system" are often thought of as only the irrigation system being designed. Upstream from a water meter, a competent irrigation designer must consider what valves are in the total system. The point from a reservoir or "modeled" reservoir to the meter occasionally requires many valves that ultimately decide the "residual" or design pressure available. Engineering requirements to provide water to a point involve fire requirements and minimum allowable pressures to a residence or commercial site. Between the reservoir or "modeled" reservoir (pumps) there may be gate valves, pressure-relief valves, pressure-reducing valves, pump valves (pressure-sustaining), flow-control valves, check valves, etc. These (and others) function similarly to the valves required to operate and maintain an irrigation system. Each valve is placed for particular reasons, some more obvious to the designer than others. The various valves implement control for a system, be it water distribution for a subdivision or for an irrigation system. If the source of water is from a lake (i.e., in a park or golf course) considerations obviously differ.

Types of valves. One must first evaluate the valving required or generally specified for an irrigation system from the point of connection (usually a water meter) to the end sprinkler head. At the meter on the upstream side is a "curb stop"; it functions as an "off-and-on" valve and is the first real control valve in a typical irrigation system. After the "curb stop," the meter is the next "valve" encountered. Although not a true valve, it measures flow, so in effect it is a control valve. This unit becomes important in the water management program.

Choice of valves depends upon the availability or unavailability of a particular water pressure. Great pressure calls for one valve; little pressure calls for another. As terrain changes and pressure increases or decreases, other valves will be required. The types of control valves utilized can directly affect the water conservation program.

At the point of connection, after determining the available water pressure, a pump may be required. If it is, there will be need for gate valves, pump control valves, etc. The pump control valve, installed between the pump and the irrigation main, is closed when the pump is not operating. When the pump is activated, the control valve opens slowly to gradually fill and pressurize the main. If a pump is not required, it is most likely that pressure regulation is required. Rarely will the available pressure meet the desired design pressure exactly. The pressure-reducing valve or combination pressure-reducing master valve or flow controls can be utilized (as required). The master valve, a standard diaphragm valve, is used to actuate or "energize" a main-line system. The time clock control opens the valve, when the selected schedule is programmed to operate, and closes when the cycle is complete, thus eliminating the possibility of any remote control valve staying open after the watering schedule and causing overwatering or erosion damage. This is generally used and recommended for slope irrigation systems. If a master valve is not required or desired, then a simple pressure-reducing valve, if required, should be specified.

Pressure-reducing valves. Pressure-reducing valves are diaphragm valves that maintain a constant pressure at their discharge side. If a pressure setting is desired to accommodate the "worst" condition, it will most likely be too great for the intermediary valves. These conditions must be reviewed to ensure that systems operating at low pressures are not getting excessive pressures affecting their intended efficiency.

Flow-control valves. If flow controls are used in lieu of pressure-reducing valves, the designer must be knowledgeable about the flow-control function. The flow-control valve allows water to pass through until the preset capacity is reached, whereupon the flow remains constant at the preset flow and absorbs the pressure differential automatically.

Once the decision has been made to utilize pressure-reducing valves, flow controls, master valves, or pumps, or to exclude any of these control systems, other decisions regarding the system come to the fore.

Backflow prevention valves. From this beginning (the point of connection), we go to the backflow preventer, positioned at a point preceding any valves within the main line (potable water source). All valves, mains, etc., preceding the backflow preventer are considered within a potable water zone. All appurtenances beyond the backflow preventer are considered "industrial" or nonpotable.

Three types of backflow preventer valves are generally used to protect the potable water supply: the atmospheric, pressure, and reduced-pressure type. Some agencies allow a double-check valve system in lieu of these. The designer must check all agency requirements for the particular locale.

When an atmospheric valve is used, the mains are considered potable to the vacuum breaker, which must be placed downstream from the valve and 12 inches above the highest sprinkler. The pressure-type requires that the unit be placed 12 inches above the highest sprinkler and that the unit be under constant pressure. The reduced-pressure unit and the double-check valve (when allowed) can be placed below the highest sprinkler head, if desired. In the pressure type and reduced-pressure type, the equipment placed downstream is considered of a nonpotable nature. Water sources using other than a potable water source are not required to use these devices (i.e., from lakes, reclaimed sources, etc.). Each pressure or reduced-pressure backflow preventer from a potable source is provided with two gate valves that are positioned within the unit for testing by agencies for proper function.

Gate/butterfly valves. Beyond the backflow preventers are a number of valves located throughout the system, including the gate valve or butterfly valve. These valves are placed strategically to provide maintenance personnel with isolation points to control and maintain the system as required.

Pressure-relief valves. Further into the system are pressure-relief valves and additional pressure-reducing valves. Pressure-relief valves are placed strategically to relieve the pressure buildup within a system due to water hammer or pressure increases due to valve-closure times, pumps, or other conditions. Pressure-relief valves discharge water to the atmosphere when the pressure within the main exceeds the design pressure. Pressure-reducing valves (as mentioned earlier) are placed within the main-line system or at the valves to provide a pressure appropriate for the function of the irrigation system. This is usually applied when the remote control valves can no longer be adjusted to control the system effectively.

Quick-coupling valves. Quick-coupling valves are placed throughout the main system to provide maintenance capabilities for washdown and emergency plant watering for initial plant establishment. These valves are generally placed at intervals within a hose reach, close to paved edges, and at other strategic locations.

Check valves. Check valves come in an assortment, such as swing-check and spring-loaded. Swing-check valves are used in relatively flat terrain, but parts of the system go uphill from the control point to prevent water from draining out of the low area. Another method of preventing drainage to the low sprinkler heads of a system is by using spring-loaded check valves, available in sizes of 1/2 inch through 1 inch, installed in individual risers or laterals. Some clients require that these valves be placed under each sprinkler to prevent erosion and to provide simplicity in construction. This would naturally vary from site to site. For very severe terrain, systems should be relatively small. Usually valving is accomplished within a 15-foot elevation differential (maximum).

Air-relief valves. Air-relief valves are installed on large main systems. The valves discharge air when the main is being filled. Vacuum-relief valves are also installed in large main systems to allow air into a main when the main breaks, thereby preventing the entire main from collapsing. The general location of air-relief valves is at the high or low point. These are points where air may become entrapped within the system and needs to be eliminated.

Remote-control valves. Last, but not least, one must discuss the remote-control valve, the system's functioning valve. This valve opens and closes as directed by the control unit. The primary remote-control valve utilized today is the electric, normally closed, solenoid-actuated valve. The two-way solenoid valve is the most common and is diaphragm-actuated. The principle of operation is simple: a diaphragm-seat arrangement with a port allows water from the upstream side of the valve to enter a pressure chamber (diaphragm chamber) and to keep the valve closed. Venting of the chamber via a solenoid depressurizes the chamber and the valve opens. To allow the valve to close slowly and to avoid water hammer, the port in the diaphragm has to be small. Because a small hole is susceptible to clogging, screens are used to protect the port. Under high, dirty water conditions, these ports can still become clogged.

Some valves with integral internal porting do not have screens for dirty water conditions; those valves are constructed from cast iron with brass trim, brass, or plastics. Plastic valves are constructed from PVC or glass-filled cycloac or nylon, or a combination using fiberglass. Plastic valves should be used under pressure recommendations of the manufacturer (generally about 100 psi maximum). Three-way solenoid valves are not as common as two-way solenoid valves, but they are available. The attempt is made to eliminate or mitigate clogging. The principle of operation is that the three-way solenoid either allows water to enter a diaphragm chamber to close the valve or to vent it. In the two-way valve, the water is always flowing through the diaphragm (pressure) chamber, and in the three-way valve it does not. Again, to make the valve close slowly, small passages are required, and the valve is still susceptible to clogging.

To choose the correct valve, you should be concerned about potential clogging. Clogged valves do not close and can cause large runoffs of water or slope failures (a good reason for the master valve). Clogging is caused by particulate matter accumulating in the water from the source, from a line breakage, or from algae from a reclaimed source. Know the mechanical features of the valve you are considering so that your intent can be specified.

When sizing the valves of a system, a general rule of thumb is that no valve is required to be line size. Pressure-reducing valves should never be line size. (Consult the manufacturer.) Valves are typically sized to create a pressure drop of approximately 5 to 10 psi for electric solenoid valves. This provides a better operating valve. "Oversizing" of valves causes seat erosion, "fluttering," and fast responses on pressure-reducing valves.

In the selection of the valves, both types and available products can be accomplished through a matrix development such as that shown under control units. In designing a large-scale project evaluate closely all requirements and address each issue. Valve selections and positioning are very important to the "water-managed" system.

Other miscellaneous considerations. The criteria for general project design development should address the following additional subjects to assure long-term continuity.

1. *Pumps.* Establish the typical requirements of pump and related equipment capabilities and function.
2. *Moisture sensors.* Establish the needs or lack of needs for the project; state the functions required.
3. *Time allocation.* Allocate time on a daily basis, considering maintenance schedules or public use, to provide the correct control system and hydraulic needs.
4. *Pipes.* Specify the types of pipes to be used for various sizes and functions, as well as the depth of cover and maximum allowable velocities and surges.
5. *Type of sprinklers.* Evaluate the type of sprinklers required for the project in a matrix format to include manufacturer's specifications versus desired function similar to that shown in controls.
6. *Sprinkler performance.* Note the typical maximum sprinkler spacings as to type and allowable pressure variations to achieve optimum distribution.
7. *Climatic conditions.* Seek data concerning local evapotranspiration information, wind, monthly historical rainfall data, temperature lows and highs, humidity data, and other information that will assist in programming the system and performing the necessary hydraulic calculations and maintenance operation guidelines.
8. *Water source.* Find out whether it is potable or nonpotable water, the delivered pressure to the site, and any other information helpful to the designer or user.
9. *Parameters for design pressures.* Substantiate the minimum design pressure allowances to be used throughout the project. An example would be to establish allowable pressure losses for meters (if used), backflow preventers (if used, valves and types), distribution mains, lateral pipes, elevation changes within a system, etc. Sites or conditions vary considerably, but this approach can serve as the foundation for the logical flow and pressure requirements for the project.
10. *Maintenance feedback.* Include all items required by maintenance for informative and decision-making reasons. Examples might be energy-use recording devices, water-consumption recording devices, operational automatic feedbacks desired, and a complete analysis of the proposed maintenance program.
11. *Miscellaneous.* Include other considerations that will ultimately affect master planning, such as pipe-sizing design parameters, sample control programming, sample calculations, checklists for designers, phasing schedules, etc.

Summary. Design criteria can range from informal for small projects to complex for large, complex projects. While some elements are well known and basic, a thorough review is necessary prior to the design process to achieve design continuity for a particular project.

Irrigation master planning

Irrigation master planning (IMP) at the inception involves total knowledge of the site and its unique characteristics. While IMP does not assure turfgrass water conservation, it does assist in overall energy conservation, saves on capital expenditures, and serves as a tool with which to manage the systems during the design process. The considerations include engineering data (including terrain), plant palette requirements, project phasing, local contractor capabilities, irrigation concepts and techniques, and an understanding of water conservation and management goals. As in any planning process, IMP requires establishing certain goals, implementing them, and revising them as necessary within the original parameters.

The irrigation master plan serves as the base for making economic decisions, provides advance information on potential engineering problems, serves as a utility plan for the engineer, graphically depicts all the "serve areas" conforming to all site constraints, and approved design criteria. The plan is used to monitor the project from inception to completion, and provides the base for discussion by the engineer, client, landscape architect, and irrigation designer.

Information depicted on IMP. It usually includes:

1. *Water meter.* If water meters are used, the location, size, and elevation are shown.
2. *"Serve areas."* Graphically depicting limits of the total area served by any one meter, the serve area acreage is also shown with upper and lower elevations served.
3. *Pressures.* The available pressure (and future pressure if it changes) is shown along with a design or regulated pressure if need be (by irrigation designer).
4. *Numbering.* Usually a master numbering system is implemented for the water meters, for convenience and communication with other consultants, when sites have multiple metering requirements in phasing sequence.
5. *Power.* Power meters are shown at their desired locations with power requirements identified.
6. *Sleeving.* All potential sleeves for pipes and wires across paved or hard surfaces are shown at their desired locations, including size, use, and quantity.
7. *Utility address.* Occasionally the address of the utility will be shown on the plan for ease in obtaining permits from the utility agency.
8. *Main line systems.* When separate irrigation mains are utilized, the size and location are depicted on the plan with supportive hydraulic calculations under separate cover.
9. *Flow demands.* Each serve area is assigned a maximum flow based upon peak use requirements (inches/week), time available to operate, size of area, sprinkler efficiency, control unit recommended, and practical knowledge of operation capabilities.
10. *Example of a master plan.* See drawing below.

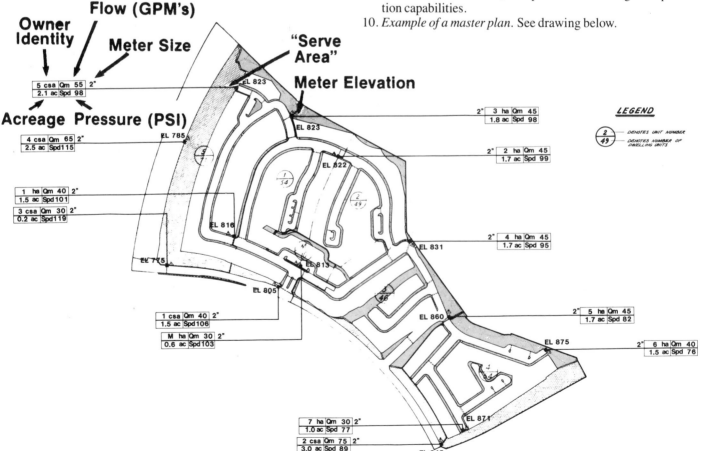

Additional considerations.

1. *Maintenance.* The maintenance staff should be consulted about concepts and preferences along with maintenance program guidelines.
2. *Economic analyses.* The irrigation master plan can serve as the document for evaluating several alternatives regarding water source, energy use, phasing economics, main-line distribution analysis, incremental or overall economic impacts, and total water consumptive options regarding plan palette selection.
3. *Monitoring.* The irrigation master plan is used to monitor all contract documents for conformance, once approved. The master plan is generally reviewed and approved by the owner, engineer, and utility companies.

"Serve areas." The "serve area" can be as small or large as all the constraints set forth by previous reasons or logical deduction. A golf course or park being served from one water source is a serve area; likewise, a small median island along a stretch of road served by a single meter or water source is a serve area. The reader may question the significance of a serve area to turfgrass water conservation. Unless one has an unlimited supply of water at any one time, it is a method to assure that the water can be delivered in the given quantity, within given time allocations, at required pressures. (There are probably other methods to determine this, but this has proved highly successful.) Without the required water availability, water conservation issues become secondary.

The serve areas discussed here are those served from domestic or reclaimed sources or any source utilizing water meters. In the design criteria section we discussed parameters for design pressures. In these parameters, certain allowable pressure losses were established on a per-meter basis. If the pressures available for main-line loss were around 10 psi, obviously the very linear serve area would be limited due to economic practicality without using excessively large pipes (i.e., square or rectangular sites could have a substantially larger acreage size than a linear landscape). Elevation differential within a serve area should be limited, due to excessive or insufficient pressure from any one source, causing pressure regulation problems and the potential of excess water usage or inability to control an area. Practical experience has indicated an optimum of 30 feet above a meter and 40 to 50 feet below a meter within the serve area. This would naturally depend on the proximity to the water source and other pressure losses that occur. Unfortunately, decisions are attained through practical experience, but it can and should be calculated for each serve area until the parameters are well understood. It is the intention to provide adequate pressure (neither excessive nor too low) within each serve area to avoid maintenance problems.

Implementation
The contract documents are prepared utilizing all information obtained and design criteria established. These documents must present to the contractor the precise information required to construct the system to be used in the water conservation program.

Design considerations. In the design criteria section it was stated that sprinkler head spacing for various sprinklers be specified. (In establishing criteria for a project, this should be specifically identified.) Spray systems are normally considered efficient at 0.6 times the diameter, while large impact or rotor systems are in the 0.5 times diameter range, depending on wind conditions. The uniformity of a particular valved system extends from consideration of elevation differential and pipe size to provide no greater than a 20 percent operating pressure differential to sprinklers with uniform application rates being valved to the same system. This offers the necessary ingredient for controlling the system from the control unit on a programmed time basis, to eliminate runoff and to provide water on a planned basis.

Examples of pipe sizing and elevation changes:

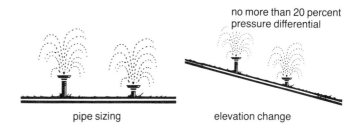

Here is an example of varying application rates:

System separation. This was stated in generalities in the design criteria section. (This should be specifically stated in project-related design criteria.) The valving for each particular valve should be established based upon exposure to the sun, soil types, plant palette, maximum allowable flow in a serve area, elevation differential, drainage patterns, and conformance to system uniformity requirements.

Types of systems. The available pressure shown in your irrigation master plan should help establish the types of sprinkler heads capable of operating efficiently in any landscaped area. Obviously, the size of the area and shape will help determine the best-suited sprinkler and placement of the right type for any given area (including subsurface irrigation). Many books are available to assist the designer in evaluating these alternatives.

The key to water conservation is uniformity, separation, and control flexibility.

Linear and looped lateral systems. Following is an example of a spray system at 15' O.C. sized in a linear and looped format. The simplicity of the looped system with all ¾-inch pipe, except for the 2-inch header, plus the 13 percent pressure differential, is worth investigating. This is shown as an idea that consultants may want to further explore for water conservation reasons (4 gpm/HD-PVC pipe).

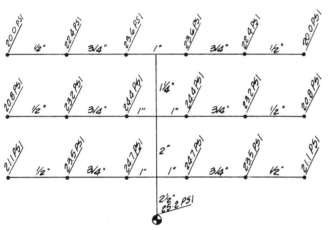

linear system
24% DIFFERENTIAL

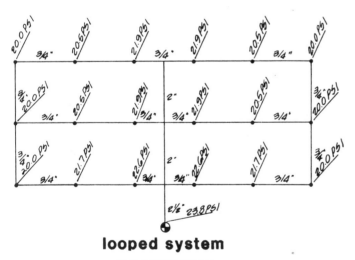

looped system
13% DIFFERENTIAL

System operation. Establishing proper operational timing for systems is a must for the betterment of plant material, less runoff and overall water conservation utilizing less duration and variable frequency of operation, depending upon soil percolation data and local evapotranspiration requirements.

Hydraulic considerations. Manufacturers provide data for sprinkler performance as gallons per minute at a given operating pressure. Note that total consumptive use and instantaneous flow for an individual system are not the multiplication of the number of sprinkler heads times the flow stated by the manufacturer. The pressure under each sprinkler head varies by the distance from the valve and losses allocated within the system. This is normally not a consideration for smaller projects, but as the project size increases from a single water source, the impact on the hydraulics of the overall system should be evaluated. The plant palette and consumptive use requirements must be evaluated and considered in the hydraulic evaluations. In the design criteria, we suggested a statement regarding acceptable velocities of water in the pipes. A normal velocity for irrigation systems has been accepted at plus or minus 5 feet per second. On large-scale pumped systems, such as golf courses and parks, it is not unusual to be in the 7-to-9-feet-per-second velocity. These, along with elevation considerations, are primary considerations in establishing appropriate hydraulic function, to ensure a system capable of being managed easily by maintenance personnel to achieve optimum water conservation results. When calculating a looped system, use one of the many sophisticated computer programs available for design and analysis.

Time requirements. It is important to establish initially the total time allocations for peak operation of the system to establish hydraulic needs. Often it is forgotten or ignored to evaluate the needs of the system at germination time. The capabilities of the control system and hydraulics should be carefully evaluated to meet the needs of the initial planting, to avoid wasted water, and to germinate the landscape properly within hydraulic and time constraints.

Field control unit placements. Control units should be located, when practical, in areas where maintenance personnel can visually check the system for proper system operation and that are economically sound.

Engineering/landscape architectural plans. The irrigation system must be closely coordinated and designed, utilizing the grading and drainage information provided by the engineer, and landscape plans provided by the landscape architect. While coordination may appear to be a quite obvious step, many plans are prepared without it. It is mentioned only as a reminder, but it can substantially affect the system's capability to meet water conservation goals, if it is not recognized.

Irrigation redos. Systems installed in the late fifties, sixties, and even the seventies are currently going through various rehabilitation steps from changing controls to complete redos. The older systems are no longer efficient for maintenance personnel for many reasons, and with the new equipment now available maintenance personnel are finding it cost effective to plan for rehabilitation.

Golf courses, once "wall-to-wall" green, are now being revised to provide complete flexibility to irrigate only those areas with high priority, if desired, in programming with computer systems. This flexibility allows the superintendent to control greens, tees, roughs, fairways, aprons, and landing areas separately, if desired. The separations assist in working more closely with maintenance and fertilizer programs, along with planned cutbacks in water use as costs of energy and water increase.

When approaching a total redo, the principles of design for new projects pretty well hold true, except that with existing conditions, more care and plan precision are required. Besides the obvious hardscape and sleeving problems, there are tree roots, existing landscape thatch buildup, utilities, and old irrigation systems. The landscape has matured, and now there may be dense groves of trees requiring a separation of systems due to the shaded area, different growing conditions, and other conditions not foreseen in the original design.

Depending on the time of year for the reconstruction, the designer must evaluate the consequences of "down time" of areas and methods to keep watering existing areas during this transition period.

Salvaging any portion of an older system must be carefully evaluated on a cost-benefit basis. Older mains may be corroded beyond another 5 years' use, valves may be marginal if repaired, and controls are very outdated. Wires in the ground, if they are chosen to be used again, must be carefully checked out.

Irrigation re-dos will be required throughout the 1980s and beyond to bring older systems up to date and to provide water managers with the tools for water conservation. The same rules apply here as in new construction—the key is proper system separation and flexibility in the method of control.

Miscellaneous. Specifications are required for bidding purposes, but can be utilized for water conservation input in such areas as metering special data, control programming information, etc.

Details for construction should be precise in establishing the desired end product to assist in water conservation.

Water management information should be provided on the contract documents regarding acreage, water pressure, each valve's hydraulic flow, and location for water management documents.

Field observation. The irrigation designer must provide field observation to assure that all design considerations are incorporated into the end product. The designer should verify the system "as built" for future water management programs, provide pressure checks on newly constructed distribution mains, and review installation of all detailed items to conform to plans and specifications. Providing the field observation and appropriate certifications will assure a truly manageable irrigation system.

Maintenance guidelines

The environment and pleasing appearance of an area are important to the development of an irrigation system. While an operations maintenance manual is generally established to focus on the irrigation and water management programs, maintenance personnel must also understand the maintenance requirements of the landscape. For example, overfertilizing can be a problem and can cause toxic levels of salt in the root zone. One can note the "proper" amount needed and that more can cause harm. It should also be noted that maintenance programs are often recommendations, and a "range" (rather than a precise definite line) is usually specified. Maintenance personnel should be evaluated to ensure reasonable knowledge and background, including familiarity with plant species, soils, turfgrass care, plant pathology, and all aspects of irrigation, generally and specifically.

It is the intention of this section to give examples of how an irrigation maintenance program *might* be formulated for maintenance staff. A maintenance manual should provide recommended water uses with methods to measure and monitor results. It is up to the managers to record and monitor total water usage while evaluating results.

Soil management involves periodic cultivation and amendments. Soil types vary in percentage of moisture held, drainage characteristics, percolation rates, tendency toward compaction, etc., and these variations greatly affect water-use efficiency and require attention, especially in low maintenance areas. In typical situations, 50 percent of the water applied per irrigation is lost to excessive percolation, runoff, or evaporation. Obviously, a good water management program must get the water where and when it is needed.

A computer or automatically controlled system does not guarantee less water use or necessarily correct water use; it is only the tool to provide this possibility.

Droughts will occur again, the continuing and future water shortages will become increasingly severe, and understanding of the watering program will become a necessity.

General goals of an irrigation maintenance program. Listed here are some goals to be considered by the water manager:

1. Water conservation is not a goal in itself, but rather a very effective tool for assuring a dependable water supply for all uses.
2. Remember the main objective of irrigation—to replace the water used in evapotranspiration as infrequently as possible.
3. The sprinkler industry will continue to change over the years, and emphasis will change from product to water management ideas. Keep abreast of changes.
4. The intent of an irrigation maintenance program is to provide insight and recommendations for the operation and maintenance for landscape water use. Additionally, experience helps one attain the ultimate maintenance requirements for the project.
5. Learn the fundamentals of a particular irrigation system, and through experience "fine-tune" the initial recommendations.
6. Establish a routine procedure of inspection and control.
7. Communicate an understanding of the basic soil-plant-water relationship and of the primary needs of turf management.
8. Establish a "management-by-the-season" program. Time of operation and duration of system operation based upon the system itself, its location in the microsystem, and climatic changes—all play an important role in management.
9. Components will be damaged or wear out; replace them in context with the original design.
10. Measure, communicate, and record water and energy use, climatic data, and reasons for changes in the program as they occur.

11. Create a sense of pride, commitment, understanding and desire to get all systems better managed and create understanding of the consequences, compromises, and benefits of good management.
12. Understand the potential risks in overwatering.

Designing for proper management. In this section is discussed factors to consider in providing a proper irrigation system and in assisting the water manager to understand the designed system. The factors mentioned are not intended to be all-inclusive.

1. Understand and apply the soil-plant-water relationship. (The mean neglect of water resources is the mismanagement of this relationship.)
2. Consider the changing evapotranspiration (ET) rates on a seasonal or possibly even a daily basis. Total water requirements equal evapotranspiration, plus leaching, minus water stored in soil reservoir, minus effective percolation.
   - Radiant energy—ET increases as radiant energy increases.
   - Temperature—ET increases as temperature increases.
   - Humidity—ET decreases as humidity increases.
   - All recommendations shown in a manual for ET are based upon historical data and do not guarantee a specific year will fall into this category.
3. The manager should investigate and evaluate the soil-moisture absorption rate.
4. Evaporation is the natural process of changing water to vapor. The rate of evaporation depends on the temperature of the water and of the air and the amount of water vapor already in the air. Dry air has a greater capacity for absorbing moisture than does moist air. Wind and temperature increases cause evaporation increase. All these influences on evapotranspiration are costly and continually affect moisture losses from plants and soils.
5. Remember: Prevailing winds, wind direction, and velocity all influence control operation.
6. Initiate management of the system to avoid application of too much water, which results in excessive percolation, runoff, or evaporation.
7. Consider that one key to proper water use is understanding that the combination of climate, plant needs (i.e., deep rooting required), and soil management can maintain the highest possible balance between the landscape quality and limited water resources.
8. Soil management involves cultivation and possible wetting agents. Soil types vary in percentage of moisture held, drainage characteristics, percolation rates, and tendencies toward compaction. Soil reservoir capacity and water quality affect the system operation. Generally, soils low in clay and organic matter have low available water-holding capabilities.
9. Individual sprinkler systems must be separated for control, based upon their individual uniqueness.
10. Each system must have a uniform distribution pattern throughout. This requires a hydraulic balance.
11. Constantly measure weather conditions and record this information for future management adjustments (i.e., cloud cover, temperatures, etc.).
12. Provide a control system capable of repeating often for variable durations and ease of change in daily operations as the season or weather conditions vary.

Turf irrigation. Irrigation practices are probably the most important factor in maintaining attractive, healthy turf. The objective is to grow good quality turf with as efficient water utilization as possible. How much water, when, and at what rate to apply it are primary considerations. Faulty water management can cause weeds, disease, soil compaction, and fertility problems. Too much or too little water are the most common causes of poor turf. In an irrigation maintenance program you should set forth recommendations that will be "fine-tuned" by the manager over periods of time.

Amount of water. When water is needed, in accordance with recommendations and eventual modifications, the intent is to apply enough at each irrigation to wet the whole root zone and connect with subsoil moisture. Continuous contact between upper and lower levels of moisture is necessary to avoid formation of a dry layer of soil that roots will not penetrate. The manager should use an auger and periodically check the soil structure noting the irrigation cycle. Timing or cultivation will modify this time eventually. The amount of water to apply at any one time depends upon how much moisture is remaining at the start of irrigation, the water-holding capacity of the soil (fine versus coarse-textured), and how well the soil drains (how fast water moves downward). This can again be checked using various methods. The amount to apply is theoretically the amount of water used by the grass and lost from the soil since the last thorough irrigation. Moisture sensors have been used successfully to establish automatically this value. Other chapters cover the particulars of various amounts of water required.

Timing irrigations. The intent is to irrigate only as frequently as water is needed rather than on a rigid schedule. Study recommendations and note physical site characteristics for watering day. (Susceptibility to disease becomes greater if a thatch or mat is present and it remains wet overnight.) Loss of water from one irrigation to the next, to be replaced, will vary according to season and climatic conditions such as temperature, humidity, wind and sunlight intensity, as already discussed. Therefore, the irrigation schedule should be reasonably flexible. Ideally, water should be applied when about 50 percent of the available water has been depleted. This may take about a week in warm weather on a medium loam soil. It is sound irrigation practice to wait until temporary wilting occurs, and it does no harm to let the top inch of soil become dry. This is covered in more detail in other chapters and should be included in your maintenance program.

Application rates. Generally, apply water only as fast as the soil can absorb it to avoid runoff and waste. (Read about soil and percolation.)

Water and fertilizer management. Moderate use of fertilizer and neither excessive nor insufficient irrigation make for the most efficient use of water (and fertilizer). Turf areas managed under minimum irrigation still need to be fertilized but should not be overfertilized. Apply enough water (and fertilizer) to maintain normal growth and color and to encourage deep rooting.

Develop a program for your maintenance instructions based upon your unique site characteristics. I recommend consulting expert soil firms.

Undesirable conditions. Undesirable conditions that affect lawn grass growth adversely have a bearing on irrigation practice and lead to extra work. These are: weed growth, disease, too-wet and too-dry areas (poor drainage or poor irrigation), soil compaction, tree-root invasion, and weak grass (shallow-root system). These are not inherent soil differences. They arise through incorrect preparation or poor maintenance generally. The causes can usually be removed or corrected, and the manager should use more efficient management practices. Some of the items that follow are covered in other chapters and are listed here only to emphasize the need to incorporate them into the maintenance program.

Excessive irrigation, either of the whole lawn or of certain parts, not only wastes water (and fertilizer), but it encourages weeds. Humps and hollows, where water accumulates and weeds persist in established turf, usually cause ponding. A uniformly dense, healthy turf is the best weed preventative.

A faulty sprinkler pattern may be due to poor servicing of sprinkler heads or initial incorrect design. Heads should be inspected regularly to make sure they are not plugged up, broken, or covered with grass.

Allowing formation of considerable thatch and mat layers not only interferes with water penetration, but when kept wet they provide a medium for the growth of harmful fungi. Evening watering should be avoided under these circumstances. The condition can be corrected with vertical mowing or aerification as required.

Mowing grass too short weakens it and commonly causes turf problems. Reducing the leaf surface exposed to light reduces the root system accordingly. Shallow-root systems require frequent watering, daily in some instances. A healthy, deep-rooted grass mowed high is less sensitive to neglect and will require less frequent irrigation and, on the whole, a saving of water.

More frequent deep watering of those areas where tree roots invade turf is necessary. This is indicated by early wilting and yellowing of the grass in streaks or patches near trees. In time, the condition should be corrected by severing the tree roots and deep watering the trees to encourage deeper roots.

Soil compaction, causing poor turf and bare spots in certain areas due to lack of soil, air, and water, is most often caused by foot traffic and travel by heavy implements or vehicles, especially when the turf is wet. Aerification measures and exclusion of all traffic from the packed soil to obtain and maintain good water penetration are required.

Soil salinity is a cause of deteriorating turf, mainly in low-lying areas. The high salt condition may be created or worsened by excess irrigation (raising the water table). It can be corrected only by leaching the topsoil, for which the water table may have to be lowered. Only salt-tolerant plants, such as some of the bermudas and tall fescue, should be planted in these problem areas.

Design criteria incorporated in the project. The imminent repetition of water shortages and the ultimate decreasing availability of this precious resource prove that a sound water management policy for landscape irrigation is a necessary and worthy endeavor. The supplemental benefits realized by reduced damage to plant material by not overwatering offer additional practicality to such conservation efforts.

The water manager should be provided with some background information on the design criteria implemented for the project. The intent should be to enlighten the manager that each portion of the design was for a particular reason and that any future changes or replacements should follow the same design philosophy.

The maintenance manual, specific and detailed, will concern:

Sprinkler heads. The sprinklers designed for this project have been spaced at 50 percent of the diameter for large rotors and 60 percent of the diameter for spray systems. This technique is a standard practice for good irrigation coverage and uniform distribution.

The nozzles selected are consistent for the various sprinklers and are key factors in calculating program times and providing uniform application rates. Repair or replacement of sprinklers over the years must maintain this same standard of design for proper irrigation practice.

Pipes. The pipes in the system are of high quality with a longevity that is most likely indefinite, unless accidental breakage or changes at the base require modification. Any future modifications should be of the same standard or higher.

The pipes have been sized to provide approximately 5 feet per second velocity as a maximum. The laterals have been sized to meet these criteria and to provide a hydraulic balance within the system of no more than 20 percent pressure differential, a normal practice of the profession.

Valves. Various valves used for the project were selected for long-term maintenance. Any replacements or future additions should be of the same quality and type.

Controls. A computer-control system with field control modules for maintenance operation and routine visual inspection has been specified. In the system programming section, the basic program schedule for the initial operation has been established. While fine-tuning will be required for the first year, no major change should occur without contacting the engineer for guidance. You may want to expand on this subject, and this is only an example. Standard controls, microprocessor, or whatever choice should be explained in some detail.

System separation. Each system has been separated as efficiently as possible, based upon landscape requirements and exposure to the sun or shaded area. This concept should always be maintained, even when changes or replacements occur.

Application rates. Each system has unique application rates based upon the discharge of water and the area covered. This varies from approximately 0.6 inches per hour to 2.8 inches per hour (discharge from the sprinkler). Each type of sprinkler has a different application rate. Due to efficiency factors, time must be adjusted to get the water to the root zones. The operational program suggested will account for these efficiency factors and exposures (if you have accounted for it).

Summary. The water manager's primary responsibility is maintaining the original design intent. While not every specific detail is utilized, the intent is to make the manager aware that all changes over the years, if required, should be made within the context of the original design. The summary should let the water manager know that all design decisions were deliberate and to change them arbitrarily would upset the water management goal.

The program schedule. On this page is a graph of a computer program depicting the operation timing of each schedule, and on page 145 is a flow chart. The schedule shown represents the irrigation system providing a net 100 percent for July. Following are recommended percent changes throughout the year.

RECOMMENDED WATER FLOW CHANGES

Month	(%)
January	0
February	0
March	13
April	47
May	80
June	93
July	100
August	93
September	70
October	50
November	18
December	0

The intention of this format is to establish a balanced program and to establish a guideline for the water manager to be able to establish percentages as the season changes.

The valve schedule, below, shows the valve numbers, group, start times, and minutes per operation suggested for peak operating times in July. Two start times per day are shown.

The flow chart, opposite page, shows the valve operation related to unit time and total flow for the sample project. This can be used for pumping analysis and peak operating evaluations for hydraulic calculations.

Example of program schedule for computer program using CRT and field station terminals

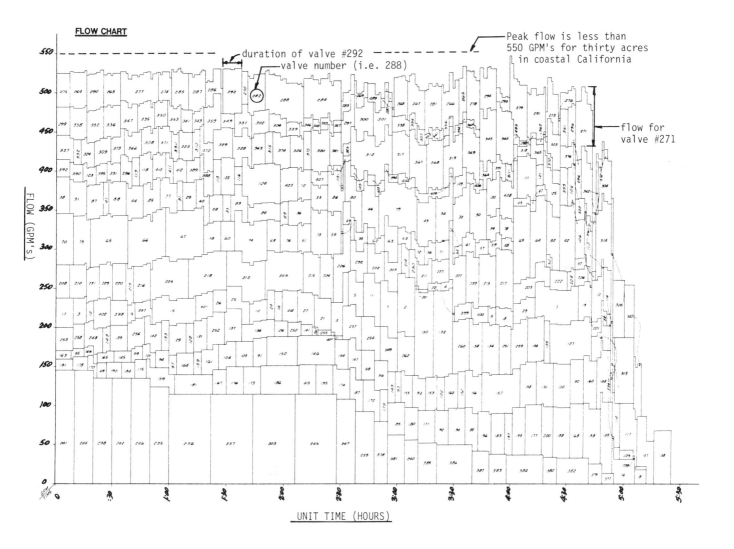

General irrigation maintenance. Many factors that are not a part of this irrigation program influence the entire maintenance program. The manager should establish a program for fertilizing, mowing, reseeding, aerification, topdressing, and dethatching. Additional programs should be established to kill weeds and crabgrass, and to prevent development of disease and insects.

Examples of routine inspections follow. Each project must be modified according to needs.

Controls and valves. The manager should inspect the controls and valves at least once a month from May to September for proper operation and system regulation. Managers should remove dust and other debris from inside the controls and clean any electrical contacts with an appropriate electrical contact cleaner. All controls should be kept pest and mildew free.

Sprinklers. The manager should keep all sprinklers clean and adjusted at least once a week during May to September and monthly for the balance of the year. Vandalism or any damage to equipment should be noted and repaired. March, normally the beginning of the season operation, always requires a very thorough review before operation.

Coverage. Once a month sprinklers should be checked for coverage to assure that they are not blocked by grass or plants. At this time, all sprinklers should be checked for damage by mowers or vehicles and repaired as necessary.

Soil. As often as necessary, the soil should be checked with an auger and evaluated with regard to operating duration and frequency. When the systems are fine-tuned, this practice should continue every other month and be evaluated, along with corrections as necessary and the keeping of records.

Computer operation. With two-way field equipment the central control unit should be checked with field valves for sequencing, timing accuracy, and general function at the beginning of each watering season, again in July, and as often as is decided necessary. Adjust this section if other controls are specified.

Records. The manager should create a form to record water usage, weather data, soil data, and system operation for perma-

TABLE 1 Recommended sample of landscape water requirements for turfgrass based upon historical data*

Month	Mean temp. (T)	Day hours (P)	Use factor (F) (T×P(%))	Use coefficient (K)	Use inches (F × K)	Ave. rain	Effect. rain	Use minus rain	Irrigation[†] requirement (inches/day)
Jan.	54.5	7.09	3.86	0.24	0.93	3.31	2.98	−1.15	nil
Feb.	55.5	6.90	3.83	0.38	1.46	3.40	2.78	−1.32	nil
Mar.	57.5	8.35	4.80	0.55	2.64	2.03	1.82	0.82	0.0265
April	60.0	8.79	5.27	0.70	3.69	0.80	0.76	2.93	0.0977
May	63.5	9.71	6.17	0.88	5.43	0.09	0.25	5.18	0.1671
June	66.5	9.69	6.44	0.92	5.92	0.06	0.06	5.86	0.1953
July	70.0	9.87	6.90	0.94	6.49	0.01	0.01	6.48	0.2090
Aug.	71.0	9.33	6.62	0.92	6.09	0.05	0.05	6.04	0.1948
Sept.	69.5	8.36	5.81	0.80	4.65	0.23	0.23	4.42	0.1473
Oct.	65.0	7.90	5.14	0.72	3.70	0.45	0.45	3.25	0.1048
Nov.	60.5	7.02	4.25	0.54	2.30	1.21	1.15	1.15	0.0383
Dec.	56.5	6.92	3.90	0.35	1.37	3.17	2.86	−1.49	nil

*This is shown for guideline use only and is to be modified accordingly as the weather changes. It is an example of the Blaney-Criddle formula applied to turfgrass water use.
[†]Excludes irrigation efficiency. This is net water required to the root zone; 1 inch of water per week = .1429 inch per day.

nent records and for use in water management decisions and fine-tuning. See Table 1, above, and Table 2 on page 147.

Miscellaneous maintenance items. Here is a brief list of maintenance problems requiring regular attention.

Backflow preventer. Regular testing, inspection, and maintenance are essential to protect the potable water supply.

Pressure vacuum breaker. Its most common problems are: leakage of water from vent area, caused by worn vent seal washer; leakage through valve seat, caused by worn valve seat washer or debris embedded in seat area (remove debris and/or replace valve seat washer); and water seepage through valve when control valve is shut off, caused by worn valve seat washer or debris embedded in seat area (remove debris and/or replace valve seat washer).

Reduced-pressure-principle backflow preventer. Periodic discharge of water from the relief valve is not a problem; the valve is performing its natural, intended function. The reduced-pressure-principle backflow preventer should be tested regularly and an RP test kit used in accordance with operating procedures.

Quick couplers. The one-piece-body type uses a U-shaped gasket to seal and prevent water from leaking around the key. This U-shaped gasket can be replaced when it becomes damaged or worn out. Should water leak around the valve key, when it is inserted in the valve, a damaged or worn-out U-shaped gasket or damage to the machined surface of the key may be responsible.

The valve also contains a rubber seal to seal off the water supply when the valve is not in use. Before removing the valve for repair, push down on the valve cage with a rod and allow water to flush away any debris that could be causing the leaks. Should this fail to stop the leaking, you will need to repair the valve.

Valve keys may start showing wear after a few years of use or they may become damaged due to improper handling. The machined surface above the steel lug or pin is the area that is used to seal. This area should be checked for wear or damage.

Piping. This section, provided to assist in making piping repairs, does not cover all situations and types of pipe, but only PVC pipes generally in use today for irrigation systems. In repairing solvent-weld PVC pipe, the basic principles of solvent cementing should be understood to assure good welded joints. In utilizing solvent-weld techniques, observe the following guidelines:

The joining surfaces of the pipe and fittings must be softened and made semifluid so that a homogeneous material bond will result. An adequate amount of cement must be applied to fill completely the void between the outside pipe wall and the fitting. An excessive amount of cement on the pipe is much better than too little. Any excess can be wiped off after the joint is put together.

The pipe must be assembled into the fitting while the surfaces are still wet and fluid so that the material will easily flow together and form a homogeneous weld or bond. The strength of a joint will increase as the cement hardens. In the tight-fitting portion of the joint, the surfaces will fuse together, while in the looser areas, the cement will tend only to bond to both surfaces. A suitable primer will usually soften more quickly and provide better penetration than just cement alone.

Impact-drive sprinklers. Any foreign object that obstructs flow through the nozzle will interfere with operation of the oscillator arm. Small rocks and pipe scale are common in some water systems. When this material is too large to pass through the nozzle's orifice, it will completely or partially plug it, resulting in a malfunction of the sprinkler rotation.

Some nozzles, when tightened back into the sprinkler, must be located in the correct position for the sprinkler to drive correctly. Even a slight variation from the correct position can affect the sprinkler's driving force and rotation.

Refer to the manufacturer's recommendations for additional information.

Spray heads. Spray heads that do not pop up require little maintenance. Most spray-head nozzles have screens that may require cleaning. If no screen is available, strainers should be specified, as the nozzles are much smaller than those noted. The nozzle opening on part-circle-pattern spray heads may become clogged or coated with a buildup of chemicals from the water supply, etc. Again, prepare this recommendation for a particular product that you have designed into the project.

Valves. The manual might include troubleshooting valves that will not open; possible valve problems (problems isolated to the valve itself); and troubleshooting valves that will not close. Consider in your maintenance program all types of valves for the project.

Control units. Before pulling a control unit for repair, be sure that the problem is not caused by a failure in another part of the automatic irrigation system. Many suspected control problems are not the result of a control failure but of a malfunctioning in some other part of the system. Ask yourself:

- Do you have 120-VAC power connected to the controller? (Someone may have turned off this power supply and failed to turn it back on.)
- Is the controller ON-OFF-AUTO switch in the AUTO position? (Perhaps the switch was turned to the OFF or MANUAL position and not returned to AUTO.)
- Have you read the manufacturer's recommendations for specific maintenance of the field unit and computer unit?

Wiring. Many times a suspected control failure will prove to be a valve or field-wire problem. When the control unit appears to operate normally and the problem is that one or more valves are not functioning correctly, develop procedures to find the problem. Develop additional procedures for locating broken or defective underground wires.

Miscellaneous. Develop any miscellaneous maintenance ideas or programs for your particular program.

Water management programs

Water management programs. In establishing a water management program, focus on the word "management." To manage effectively, one needs to have numbers, goals, and a comprehensive knowledge of management's function.

Measurement. For your management program choose water meters or flow meters to measure against established goals and to evaluate performance. As in irrigation management systems, you will find that most recommendations are just that—recommendations or guidelines. To manage the system, you will need knowledgeable people, sound initial design criteria, firm implementation procedures, and well-documented, recommended water-usage requirements. To establish these numbers, month-to-month irrigation recommendations will be required. The average rainfall, ET, and other data will be required to formulate these charts that are converted to water meter units. An example is Table 2.

TABLE 2   Rainfall data and irrigation recommendations

			TURF		SLOPES/GROUNDCOVER	
MONTH	AVERAGE RAINFALL (Acre - Inches)	EVAPOTRANS-PIRATION RATE FOR TURF (Acre - Inches)	RECOMMENDED ARTIFICIAL IRRIGATION (Acre - Inches)	TOTAL RECOMMENDED ARTIFICIAL IRRIGATION (100 Cu.Ft./Acre)	RECOMMENDED ARTIFICIAL IRRIGATION (Acre - Inches)	TOTAL RECOMMENDED ARTIFICIAL IRRIGATION (100 Cu.Ft./Acre)
January	2.75	0.85	0.25	9.1	0.20	7.3
February	3.10	1.00	0.35	12.7	0.25	9.1
March	2.60	1.85	0.60	21.8	0.45	16.3
April	1.10	2.60	1.50	54.5	1.15	41.8
May	0.40	3.60	3.20	116.2	2.40	87.1
June	0.05	4.50	4.45	161.5	3.30	119.8
July	0.00	5.05	5.05	183.3	3.75	136.1
August	0.05	4.90	4.85	176.1	3.10	112.5
September	0.15	3.90	3.75	136.1	2.80	101.6
October	0.70	2.70	2.00	72.6	1.50	54.5
November	1.25	1.60	0.50	18.2	0.35	12.7
December	2.60	1.00	0.35	12.7	0.25	9.1
Yearly Total	14.75	33.55	26.85	974.8	19.50	707.9

The areas shown shaded theoretically do not require irrigation. We have shown a 30% value of evapotranspiration for budgeting as watering to some degree will be required. December through February will require one fifth of the April setting normally and up to two-thirds in unusual hot weather. March and November will require roughly one-third to the full April reading and will have to be adjusted as weather dictates.

Metered areas. Using the unit information established previously, apply the multiplication of acres in the "serve area" to the units and establish expected quantities per given metered area. The following example can also be used as a monitoring form:

Establishing a water budget. Using the information on total consumptive use for a given area, yearly and monthly use can be projected. Using the water rates from the purveyor, one can simply establish the budget based upon expected use. Use for any one year can be different from that expected, and, therefore, allowances should be made for potential adverse conditions.

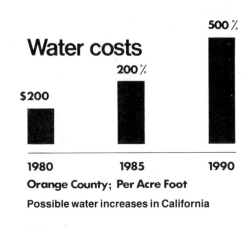

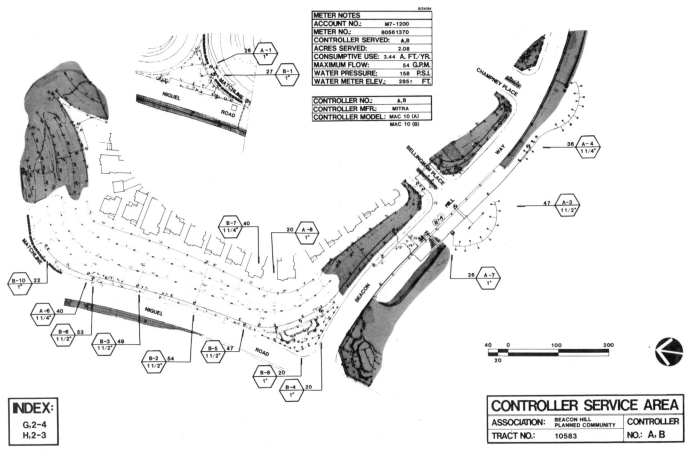

An identification map (park with three controls)

Variations in use. Even weather forecasters with sophisticated satellite control have difficulty predicting the weather, so we cannot recommend establishing a rigid chart. As stated previously, the system is in a constant state of adjustment because of climatic changes, and the manager must be aware of these conditions and what to do in each case. The water manager will be required to react quickly to these changes.

Monitoring a system. In a water management program, "user" charts are provided the owner to record water consumed versus recommended usage. Careful records offer the manager useful information for future water budgeting and management.

Assisting in guidelines of programming. The irrigation designer should supply individual valve scheduling, showing timing per day, day-per-week scheduling, and the relationship to total consumptive use so that the water manager can fine-tune the operation.

Example of a conventional programming method. The following is an example of one method of providing a conventional control program schedule. A computer method is shown in the maintenance section. This programming should coincide with overall consumption use graphs.

Example of a typical monitoring form

Education of the "user." A key objective of a water management program is, obviously, to gain beneficial results. Educating the owner and maintenance personnel regarding the details of the proposed management program has proved very effective. This generally takes place in the classroom and the field; with experienced operators, a general understanding of the goals and objectives can take place in 4 to 8 hours. Periodic followups and evaluation are vital to continuation of the program.

Benefits. Among the more obvious benefits of a water management program are savings in water cost, total controlled water use, and potential energy savings. The other benefits apply to better turfgrass and general plant material growth.

Conclusion

Many items come into consideration when establishing a truly manageable system that have been discussed in other sections of this chapter and throughout the entire text. There are many other methods of establishing water management programs; each should be considered, based upon maintenance personnel capabilities and simplicity of effective monitoring. Grading, drainage, and other considerations cannot be overlooked, and the effects of these elements must be made clear to the manager. The best method is obviously the one that suits the user, but each one should go through a sound design criteria evaluation that establishes the foundation for the irrigation system engineering.

Glossary

absorption rate—Rate at which the soil will absorb water. It is not a static rate as it includes the infiltration rate and the infiltration capacity of the soil.

aeration, mechanical—See **cultivation, turf**.

aerify—See **cultivation, turf**.

amendment, physical—Any substance such as sand, calcined clay, peat, and sawdust, added to soil to alter physical conditions

angle valve—Valve in which water enters in plane 90 degrees from the plane at which it exits

aquifer—An underground formation that contains sufficient saturated permeable material to yield significant quantities of water to wells and springs; a reservoir for underground water.

arch (or a **sprinkler head**)—Throw of a head from side to side, usually expressed in degrees and measured looking down on the head

artificial turf—Synthetic surface simulating turf

ball mark—A depression and/or tear in the surface of a turf, usually a green, made by the impact of a ball

ball roll—Distance a ball moves (a) after striking the ground upon termination of its air flight, (b) as the result of a putting stroke, or (c) as a result of hand-imparted motion as in lawn bowling

bed knife—Stationary bottom blade of a reel mower against which the reel blades turn to produce a shearing cut. The bed knife is carried in the mower frame at a fixed distance from the reel axis and an adjustable fixed distance above the plane of travel.

bench setting—Height a mower's cutting plane (bed knife or rotating blade tip) is set above a hard, level surface

blend, seed—Combination of two or more cultivars of a single species

bgd—Billions of gallons per day

block or **battery** (of head)—Group of heads controlled by one valve

bolson (Spanish)—Closed or confined flat-floored desert valley, which drains to a playa or, if underground, to an aquifer

broadcast sprigging—Vegetative turf establishment by broadcasting stolons, rhizomes, or tillers and covering with soil

brushing—Practice of mowing a brush against the surface of a turf to lift nonvertical stolons and/or leaves before mowing to produce a uniform surface of erect leaves

bunchgrass—See **bunch-type growth**.

bunch-type growth—Plant development by intravaginal tillering at or near the soil surface without the formation of rhizomes or stolons

bushing down—Making a change in pipe size from larger to smaller

bushing up—Making a change in pipe size from smaller or larger, usually to accommodate a fitting or other component

calcined clay—Clay minerals, such as montmorillonite and attapulgite, that have been fired at high temperatures to obtain absorbent, stable, granular particles; used as amendments in soil modification.

castings, earthworm (wormcasts)—Soil and plant remains excreted and deposited by earthworms in or on the turf surface or in their burrows. The formation of these relatively stable soil granules can be objectionable on closely mowed turf by producing an uneven surface.

catcher—Detachable enclosure on a mower used to collect clippings; also called basket, bag, or box.

centrifugal spreader—Applicator from which dry, particulate material is broadcast as it drops onto a spinning disc or blade beneath the hopper

chemical trimming—Using herbicides or chemical-growth regulators to limit turfgrass growth around trees, borders, monuments, walks, etc.

class (of pipe)—Designation given to pipe, usually expressed in maximum pressure for which the pipe should be used

cleavage, plane, sod—Zone of potential separation at the interface between the underlying soil and an upper soil layer adhering to transplanted sod. Such separation is most commonly a problem when soils of different textures are placed one over another.

clipping removal—Collecting leaves cut by mowing and removing them from the turf

clippings—Leaves and stems cut off by mowing

clonal planting—Vegetative establishment using plants of a single genotype placed at a spacing of 1 meter or more

coefficient of uniformity—Percentage figure derived from the precipitation rates at various points of a sprinkler system; used in technical circles to determine the efficiency of a sprinkler system.

cold water insoluble nitrogen (**WIN**)—Form of fertilizer nitrogen not soluble in cold water (25°C)

cold water soluble nitrogen (**WSN**)—Form of fertilizer nitrogen soluble in cold water (25°C)

colorant—Dye, pigment, or paintlike material applied to turf to create a favorable green color when the grass is discolored or damaged

combing—Using a comb, with metal teeth or flexible tines, fastened immediately in front of a reel mower to lift stolons and procumbant shoots so that they can be cut by the mower

consumption—Portion of water withdrawn for offstream uses and not returned to a surface or groundwater source. In plants and animals the water is used metabolically.

controlled-release fertilizer—See **slow-release fertilizer**.

controllers or **programmers**—1. Clock
    a. Hydraulic
        (1) tubing control to valve
    b. Electric
        (1) wire control to valves
    c. Nonvariable
    d. Variable
    e. Infinitely variable
2. Manual
    a. Remote control
    b. Spring-loaded

cool-season turfgrass—Species best adapted to growth during cool, moist periods of the year; commonly having temperature optimums of 15° to 24°C; e.g., bentgrasses, bluegrasses, fescues, and ryegrasses.

coring—Method of turf cultivation in which soil cores are removed by hollow tines or spoons

coupler key—Hollow shaft used to connect the sprinkler head to the quick coupler valve. Water passes through the key to the head.

cover, winter-protection—See **winter-protection cover**.

coverage—Refers to the way water is applied to an area. Coverage can be in relation to the throw of a head against the spacing of it or the overall job the head or system is doing in irrigating the turf.

creeping growth habit—Plant development by extravaginal stem growth at or near the soil surface with lateral spreading by rhizomes and/or stolons.

cultipacker seeder—Mechanical seeder designed to place turfgrass seeds in a prepared seedbed at a shallow soil depth followed by firming of the soil around the seed. It usually consists of a pull-type tractor rear-mounted unit having a seed box positioned between the larger, ridged, front roller and an offset, smaller, rear roller.

cultivation, turf—Applied to turf, refers to working of the soil without destruction of the turf, e.g., coring, slicing, grooving, forking, shattering, spiking, or other means

cup cutter—Hollow cylinder with a sharpened lower edge used to cut the hole for a cup in a green or to replace small spots of damaged turf

cushion—See **resiliency**.

cutting height—Of a mower, the distance between the plane of travel (base of wheel, roller, or skid) and the parallel plane of cut

density, shoot—See **shoot density**.

dethatch—Procedure of removing an excessive thatch accumulation either (a) mechanically as by vertical cutting or (b) biologically as by topdressing with soil

detritus—Loose unconsolidated fragments of rock, minerals, and soil particles resulting from erosion

distribution—Refers to the way sprinkler head(s) apply water to the turf

distribution curve (of a sprinkler head)—Plotted curve of the precipitation rate of an individual head at various points on its radius

divot—Small piece of turf severed from the soil by a golf club or the twisting-turning action of a cleated shoe

dormant seeding—Planting seed during late fall or early winter after temperatures become too low for seed germination to occur until the following spring

dormant sodding—Transplanting sod during late fall or early winter after temperatures become too low for shoot growth and rapid rooting

dormant turf—Turfs that have temporarily ceased shoot growth as a result of extended drought, heat, or cold stress

dry spot—See **localized dry spot**.

effective cutting height—Height of the cutting plane above the soil surface at which the turf is mowed

establishment, turf—Root and shoot growth following seed germination or vegetative planting needed to form a mature, stable turf

evaporation—Water evaporated as vapor from land, water, and vegetation surfaces

evapotranspiration—Water vapor evaporated from surfaces plus that transpired by plants

fertigation—Application of fertilizer through an irrigation system

fertilizer burn—See **foliar burn**.

flail mower—Mower that cuts turf by impact of free-swinging blades rotating in a vertical cutting plane relative to the turf surface. See also **impact mowing**.

flow—Movement of water expressed in gallons per minute (GPM)

foliar burn—Injury to shoot tissue caused by dehydration due to contact with high concentrations of chemicals, e.g., certain fertilizers and pesticides

foot-head—Measurement of pressure based on the fact that a column of water 1 foot high has a 1 foot-head rating due to the weight of the water. One foot-head is equal to .433 pounds per square inch (psi).

footprinting, frost—Discolored areas of dead leaf tissue in the shape of foot impressions that develop after walking on live, frosted turfgrass leaves

footprinting, wilt—Temporary foot impressions left in a turf when flaccid leaves of grass plants suffer incipient wilt and have insufficient turgor to spring back after treading has occurred

forking—Method of turf cultivation in which a spading fork or similar solid tine device is used to make holes in the soil

french drain—See **slit trench drain**.

frequency of clip—Distance of forward travel between successive cuts of mower blades

friction loss—Amount of pressure loss due to friction in pipe, valves, or other components of a water system incurred with the movement of water in or through those components

functional water use—Category of offstream use of water, e.g., domestic, commercial, manufacturing, agriculture, steam electric generation, minerals industry

GPM—Measure of water flow (gallons per minute)

gate valve—Low-friction loss valve used to manually control the flow of water. Sometimes used as a throttle valve to induce a restriction in the water line which causes a pressure loss under flow conditions.

globe valve—Valve in which water enters and exits in the same plane

grading—Establishing surface soil elevations and contours before planting

grain, turf—Undesirable, procumbently oriented growth of grass leaves, shoots, and stolons on greens; a rolling ball tends to be deflected from a true course in the direction of orientation.

gravity flow—Movement of water due to elevation differences. Water is often transferred from one place to another this way as no pumps are necessary.

grooving—Turf cultivation method in which vertical, rotating blades cut continuous slits through the turf and into the soil, with soil, thatch, and green plant material being displaced

groundwater—Water located below the water table and contained in aquifers; usually does not extend below 2,000 to 3,000 feet.

head-to-head spacing—Locating of heads so that the throw of one head will reach the other head, giving 100 percent overlap

hole punching—See **cultivation, turf**.

"hot" line—Water line that is under pressure at all times

"hot" soil—Soil containing a high degree of chemicals that will cause rapid deterioration of metallic pipes and fittings

hot water insoluble nitrogen (HWIN)—Form of fertilizer nitrogen not soluble in hot water (100°C); used to determine the activity index of ureaforms. See **nitrogen activity index**.

hydraulic seeding—See **hydroseeding**.

hydroplanting—Planting vegetative propagules (e.g., stolons) in a water mixture by pumping through a nozzle which sprays the mixture onto the plant bed. The water-propagule mixture may also contain fertilizer and a mulch.

hydroseeding—Planting seed in a water mixture by pumping through a nozzle which sprays the mixture onto a seedbed. The water mixture may also contain fertilizer and a mulch.

impact mowing—Mowing in which the inertia of the grass blade resists the impact of a rapidly moving blade and is cut; this is characteristic of rotary and vertical mowers and in contrast to the shearing cut of reel and sickle bar mowers.

insert fittings—Type of fittings normally used with polyethylene pipe which slip into the pipe causing a restriction to the flow of water

installation cost—Cost of installing a system, including labor, overhead, and profit

installed cost—Cost of a system completely installed, including all materials plus installation cost

interseeding—Seeding between sod plugs, sod strips, rows of sprigs, or stolons

irrigation, automatic—Water application system in which valves are automatically activated, either hydraulically or electrically, at times preset on a controller. System may or may not be integrated with an automatic sensing unit.

irrigation, manual—Water application using hand-set and hand-valved equipment

irrigation, semiautomatic—Water application system in which valves respond directly to a manually operated remote-control switch

irrigation, subsurface—Application of water below the soil surface by injection or by manipulation of the water table

knitting—See **sod knitting**.

land rolling—See **roller, water ballast**.

lapping, mower (backlapping)—Backward turning of the reel against the bed knife while a fluid-dispersed grinding compound is applied. Lapping hones the cutting faces and mates the reel and bed knife to a precise fit for quality mowing.

lateral—Any pipeline that takes off from the main line of a system; generally, nonpressure lines, downstream of the valve. Synonym: branch.

lateral shoot—Shoots originating from vegetative buds in the axils of leaves or from the nodes of stems, rhizomes, or stolons

lawn—Closely mowed ground cover, usually grass

lawngrass—See **turfgrass**.

layering, soil—Stratification within a soil profile, which may affect conductivity and retention of water, soil aeration, and rooting; can be due to construction design, topdressing with different-textured amendments, inadequate on-site mixing of soil amendments, or blowing and washing of sand or soil.

leaf mulcher—Machine that lifts leaves from a turf and shreds them small enough to fall down within the turfgrass canopy

liquid fertilization—Method of fluid nutrient application in which dissolved fertilizer is applied as a solution

localized dry spot—Dry spot of turf and soil surrounded by more moist conditions, which resists rewetting by normal irrigation or rainfall; is often associated with thatch, fungal activity, shallow soil over buried material, compacted soil, or elevated sites in terrain.

loop—"Endless" connection of pipe in which water may flow more than one way to reach the point at which it is to be used

low temperature discoloration—Loss of chlorophyll and associated green color which occurs in turfgrasses under low temperature stress

main—Generally, the pressure or "hot" line pipe in a system

maintenance, turf—See **turfgrass culture**.

mat—Thatch which has been intermixed with mineral matter that develops between the zone of green vegetation and the original soil surface; commonly associated with greens that have been topdressed.

matting—Dragging steel door matting over the turf surface to work in topdressing and smooth the surface; also used to break up and work in soil cores lifted out by coring or grooving.

mgd—Millions of gallons per day

mixture, seed—Combination of two or more species

monostand—Turfgrass community composed of one cultivar

mowing frequency—Number of mowings per unit of time, expressed as mowings per week; or the interval in days between one mowing and the next.

mowing height—See **cutting height**.

mowing pattern—The orientations of travel while mowing turf. Patterns may be regularly changed to distribute wear and compaction, to aid in grain control, and to create visually aesthetic striping effects.

mulch blower—Machine using forced air to distribute particles of mulch over newly seeded sites

nitrogen activity index (AI)—Applied to ureaformaldehyde compounds and mixtures containing such compounds; the AI is the percentage of cold water insoluble nitrogen that is soluble in hot water.

$$AI = \frac{\%\,WIN - \%\,HWIN \times 100}{\%\,WIN}$$

nonpressure lines (or pipe)—Lines that are not under continual pressure and that are downstream of the valve

nursegrass—See **temporary grass**.

nursery, stolon—See **stolon nursery**.

nursery, turfgrass—Area where turfgrasses are propagated for vegetative increase to provide a source of stolons, sprigs, or sod for vegetative planting

off-site mixing—Mixing soil and amendments for soil modification at a place other than the planting site

operating cost—Cost of operating a system; includes cost of water, labor in operating the system, and maintenance of the system

operating cycle—Time from when the controller(s) starts operating the system to the time when the controller(s) halts the operation; total elapsed time.

operating pressure—Pressure at a given point in the system when it is in operation; usually taken at the base of the head or at the nozzle.

overlap—Radius of a sprinkler head in relation to the distance between heads, expressed in percentage

overseeding—Seeding into an existing turf. See also **winter overseeding**.

PSI—Measure of pressure (pounds per square inch)

pegging sod—Use of pegs to hold sod in place on slopes and waterways until transplant rooting occurs

planting bed—Soil area prepared for vegetative propagation or seed germination and establishment of turf

playa—Shallow temporary lake overlying flat-floored basin. It fills during prolonged or excessive periods of rainfall; dries up during drought.

plugging—Vegetative propagation of turfgrasses by plugs or small pieces of sod to establish vegetatively propagated turfgrasses as well as to repair damaged areas

poling—Using a long (bamboo) switch or pole to remove dew and exudations from turf by switching the pole in an arc while in contact with the turf surface; also used to break up clumps of clippings and earthworm casts.

polystand—Turfgrass community composed of two or more cultivars and/or species

precipitation rate—Rate at which water is applied in a pattern of heads; usually expressed in inches per hour.

pregerminated seed—Preconditioning seed before planting by placing in a moist, oxygenated environment at optimum temperatures to favor more rapid germination after seed

press rolling—Mechanical planter designed to push sprigs or stolons into the soil, followed by firming of the soil around the vegetative propagules

pressure—Measure of the relative force of water expressed in pounds per square inch (psi) or feet-head

pressure drop—Drop of pressure due to friction or restrictions when water is in motion or due to an elevation rise

pressure lines (or pipe)—Lines that are under continual pressure

pseudo thatch—Upper surface layer above a thatch; composed of relatively undecomposed leaf remnants and clippings.

puffiness—Spongelike condition of turf that results in an irregular surface

quick-coupler valve—Valve located off pressure lines that can be opened to a surface outlet by use of a coupler key

rating (of pipe)—Pressure rating given to pipe to denote the maximum pressure for which the pipe should be used

rebuilding—Practices that result in complete change of a turf area

recuperative potential—Ability of turfgrasses to recover from injury

reel mower—Mower that cuts turf by means of a rotating reel of helical blades which pass across a stationary blade (bed knife) fixed to the mower frame; this action gives a shearing type of cut.

reestablishment, turf—Procedure involving (a) complete turf removal, (b) soil tillage, and (c) seeding or vegetative establishment of a new turf; does not encompass rebuilding.

release rate, fertilizer—Rate of nutrient release following fertilizer application. Water-soluble fertilizers are termed fast-release, while insoluble or coated soluble fertilizers are termed slow-release.

remote-control unit—Manually operated unit at which remote control valves, regardless of their location, can be operated

remote-control valve—Valve, operated either electrically or hydraulically, that can be operated from a point distant from its location

renovation, turf—Improvement usually involving weed control and replanting into existing live and/or dead vegetation; does not encompass reestablishment.

reseeding, turf—To seed again, usually soon after an initial seeding has failed to achieve satisfactory establishment

residual response, fertilizer—Delayed or continued turfgrass response to slow-release fertilizers, lasting longer than the usual response from water-soluble fertilizers

resiliency—Capability of a turf to spring back when balls, shoes, or other objects strike the surface, thus providing a cushioning effect

rippling—Wave or washboard pattern on surface of mowed grass, usually resulting from mower maladjustment, too fast a rate of mower travel, or too low a frequency of clip for the cutting height

roller, water ballast—Hollow, cylindrical body, the weight of which can be varied by the amount of water added; used for leveling, smoothing, and firming soil.

root pruning, tree—Judicious cutting of tree roots to reduce their competition with an associated turf

rotary mower—Powered mower that cuts turf by high-speed impact of a blade or blades rotating in a horizontal cutting plane.

row sprigging—Planting of sprigs in rows or furrows

saddle—Type of fitting encircling the pipe; a hole is usually burned or drilled through the pipe it encircles to furnish water through it.

scald, turf—Injury to shoots which collapse and turn brown when intense sunlight heats relatively shallow standing water to lethal temperatures

scalping—Removal of an excessive quantity of green shoots at any one mowing, resulting in a stubbly, brown appearance caused by exposed stems, stolons, and dead leaves

scarifying, turf—Breaking up and loosening the surface. See **vertical cutter**.

schedule (of pipe)—Rating similar to class but relating more to the wall thickness of the pipe; the pressures which a schedule of pipe can operate on vary with pipe size.

scorching—See **scald, turf**.

scum—Layer of algae on the soil surface of thin turf; drying can produce a somewhat impervious layer that impairs subsequent shoot emergence.

section (noun)—A group of heads and valve(s) that operate on one station of a controller or at one time on a manual system

section (verb)—Act of sectioning a system in the design stage

sectioning—Act of placing heads in sections in design work

seed mat—Fabricated mat with seed (and possibly fertilizer) applied to one side; the mat serves as the vehicle to (a) apply seed (and fertilizer), (b) control erosion, and (c) provide a favorable microenvironment for seed germination and establishment.

seeding, dormant—See **dormant seeding**.

semiarid turfgrass—Turfgrass species adapted to grow and persist in semiarid regions without irrigation, such as buffalograss, blue grama, and sideoats grama

settling, soil—Lowering of the soil surface previously loosened by tillage or by excavation and refilling; occurs naturally in time, and can be accelerated mechanically by tamping, rolling, cultipacking, or watering.

shattering—Turf cultivation method involving fragmentation of a rigid or brittle soil mass usually by a vibrating mechanical mole device

shaving, turf—Cutting and removal of all verdure, thatch, and excess mat by means of a sod cutter followed by turfgrass regrowth from underground lateral stems; used on bowling greens, especially bermudagrass.

shoot density—Number of shoots per unit area

short-lived perennial—Turfgrasses normally expected to live only 2 to 4 years

sickle-bar mower—Mower that cuts grass by means of horizontal, rapidly oscillating blades which shear the gathered grass against stationary blades

slicing—Turf cultivation method in which vertically rotating, flat blades slice intermittently through the turf and the soil

slip fittings—Fitting for plastic pipe into which the pipe fits and is glued with a solvent. There are various combinations of slip and threaded outlets on fittings for plastic pipe.

slit trench drain—Narrow trench (usually 5 to 10 cm wide) backfilled to the surface with a material, such as sand, gravel, or crushed rock, to facilitate surface or subsurface drainage

slow-release fertilizer—Designates a rate of dissolution less than is obtained for completely water-soluble fertilizers; may involve compounds which dissolve slowly, materials that must be decomposed by microbial activity, or soluble compounds coated with substances highly impervious to water.

sod—Plugs, squares, or strips of turfgrass with adhering soil to be used in vegetative planting

sod cutter—A device to sever turf from the ground; the length and thickness of the sod being cut is adjustable.

sod cutting—See **sod harvesting**.

sod harvesting—Mechanical cutting of sod, for sale and/or transfer to a planting site, with a minimum of soil to facilitate ease of handling and rooting

sod heating—Heat accumulation in tightly stacked sod; may reach lethal temperatures.

sod knitting—Sod rooting to the extent that newly transplanted sod is held firmly in place

sod production—Culture of turf to a quality and maturity that allows harvesting and transplanting

sod rooting—Growth of new roots into the underlying soil from nodes in the sod

sod strength—Relative ability of sod to resist tearing during harvesting, handling, and transplanting; in research, the mechanical force (kg) required to tear apart a sod when subjected to a uniformly applied force.

sod transplanting—Transfer to and planting of sod on a new turf area

sodding—Planting turf by laying sod

sodding, dormant—See **dormant sodding**.

soil heating—See **soil warming**.

soil mix—Prepared mixture used as a growth medium for turfgrass

soil modification—Alteration of soil characteristics by adding soil amendments; commonly used to improve physical conditions of turf soils.

soil probe—Soil sampling tool usually having a hollow cylinder with a cutting edge at the lower end

soil screen—Screen used to remove clods, coarse fragments, and trash from the soil; may be stationary, oscillating, or, in the case of cylindrical screens, rotating.

soil shredder—Machine that crushes or pulverizes large soil aggregates and clods to facilitate uniform soil mixing and topdressing application

soil warming—Artificial heating of turf from below the surface, usually by electrical means, to prevent soil freezing and maintain a green turf during winter

soiling—See **topdressing**.

solid sodding—See **sodding**.

spiking—Turf cultivation method in which solid tines or flat, pointed blades penetrate the turf and soil surface to a shallow depth

spongy turf—See **puffiness**.

spoon, coring—Turf cultivation method involving curved, hollow, spoonlike tines that remove small soil cores and leave openings in the sod

spot seeding—Seeding of small, usually barren or sparsely covered areas within established turf

spot sodding—Repair of small areas of damaged turf using plugs or small pieces of sod

sprig—Stolon, rhizome, tiller, or combination used to establish turf

sprigging—Vegetative planting by placing sprigs in furrows or small holes

spring greenup—Initial seasonal appearance of green shoots as spring temperature and moisture conditions become favorable, thus breaking winter dormancy

sprinkler heads—1. Lawn spray
    a. Surface or stationary—a fixed or stationary lawn sprinkler, as opposed to pop-up type
    b. Pop-up—sprinkler which raises the nozzle above the surface of the surrounding grass to avoid interference
2. Shrub
    a. Spray or stream—sprinkler used to water all planted shrub area as opposed to bed-spray sprinkler
3. Bed-spray—sprinkler used in small restricted bed areas and narrow planter boxes for dispersing water better than a bubbler under heavy foliage
4. Bubbler—spider stream or surface; flood—sprinkler designed to soak restricted areas
5. Rotary
    a. Pop-up—relatively large area sprinkler with one or more nozzles, which rises above the ground and rotates throwing streams of spray over a circular area driven by ball, cam, gear, or impact
    b. Aboveground—relatively large area sprinkler with one or more nozzles mounted aboveground and driven for the most part by impact

spudding—Removal of individual weedy plants with a small spadelike tool which severs the root deep in the soil so that the weed can be lifted from the turf manually

static pressure—Water pressure on a system when there is no water flowing through the system, thus incurring no friction losses

station—Refers to the station on the controller which controls a group of heads and valve(s) or heads controlled by that station

stolon nursery—Area used for producing stolons for propagation

stolonize—Vegetative planting by broadcasting stolons over a prepared soil and covering by topdressing or press rolling

strip sodding—Laying of sod strips spaced at intervals, usually across a slope. Turf establishment depends on spreading of the grass to form a complete cover; sometimes the area between the strips is interseeded.

subgrade—Soil elevation constructed at a sufficient depth below the final grade to allow for the desired thickness of topsoil, root-zone mix, or other material

summer dormancy—Cessation of growth and subsequent death of leaves of perennial plants due to heat and/or moisture stress

sump—Pit utilized for draining. In sprinkler work, a gravel-filled pit into which drain valves empty water from the sprinkler lines.

surge—Pressure increases caused by the velocity of flow of water and speed with which the flow is stopped

swing joint—Piping method used when making a connection off a line so the line will be protected if the piping leading from it is placed under strain

synthetic turf—See **artificial turf**.

syringing—Spraying turf with small amounts of water to: (a) dissipate accumulated energy in the leaves by evaporating free surface water, (b) prevent or correct a leaf-water deficit, particularly wilt, and (c) remove dew, frost, and exudates from the turf surface

tapped coupling—Fitting, usually of cast iron, for use on asbestos-cement or cast iron pipe which couples the pipe together and furnishes an outlet which is tapped to pipe size so piping can connect into it

temporary grass—Grass species not expected to persist in a turf and thus used as temporary cover

texture—In turf, refers to the composite leaf width, taper, and arrangement

thatch—Loose, intermingled, organic layer of dead and living shoots, stems, and roots that develops between the zone of green vegetation and the soil surface

thatch control—Preventing excessive thatch accumulation by cultural manipulation and/or reducing excess thatch by mechanical or biological means

threaded fitting—Plastic or steel fitting that has threaded fittings on all openings; threads are standard iron pipe size threads.

tipburn—Leaf-tip necrosis resulting from lethal internal water stress caused by desiccation, salt, or pesticide accumulation

topdressing—A prepared soil mix added to the turf surface and worked in by brushing, matting, raking, and/or irrigating (a) to smooth a green surface, (b) to firm a turf by working soil in among stolons and thatch forming materials, (c) to enhance thatch decomposition, and (d) to cover stolons or sprigs during vegetative planting; also the act of applying topdressing materials to turf.

topsoil planting—Modification of stolonizing which involves covering the area with soil containing viable rhizomes and/or stolons to establish a turf cover

transite pipe—Trade-named asbestos-cement type of pipe often used as a general term for asbestos-cement pipe

transitional climatic zone—Suboptimal zone between the cool and warm climates where both warm- and cool-season grasses can be grown

transpiration—Water transpired by plants as vapor

trimming—Cutting edges and borders of turf to form clearly defined lines

turf—Covering of mowed vegetation, usually a turfgrass, growing intimately with an upper soil stratum of intermingled roots and stems

turfgrass—A species or cultivar of grass, usually of spreading habit, that is maintained as a mowed turf

turfgrass color—Composite visual color of a turfgrass community perceived by the human eye

turfgrass community—Aggregation of individual turfgrass plants that have mutual relationships with the environment as well as among the individual plants

turfgrass culture—Composite cultural practices involved in growing turfgrasses for lawns, greens, sports facilities, and roadsides

turfgrass management—Development of turf standards and goals that are achieved by planning and directing labor, capital, and equipment; the objective is to achieve those standards and goals by manipulating cultural practices.

turfgrass quality—Composite, subjective visual assessment of the degree to which a turf conforms to an agreed standard of uniformity, density, texture, growth habit, smoothness, and color

turfgrass uniformity—Visual assessment of the degree to which a turfgrass community is free from variations in color, density, and texture across the surface

type (of pipe)—Designation denoting the material construction of the pipe

ureaformaldehyde (UF)—Synthetic slow-release nitrogen fertilizer known under the generic name, ureaform, and consisting mainly of methylene urea polymers of different lengths and solubilities; formed by reacting urea and formaldehyde.

vacuum breaker—Type of antisyphon device, either atmospheric or pressure type

vadose—Water in the earth's crust above the water table

valves—Wide assortment of devices for controlling a liquid:
1. Remote-control
   a. Hydraulic
   b. Electric
   c. Thermal
2. Manual
   a. Angle
   b. Globe
   c. Gate
3. Check
   a. Swing
   b. Low-pressure flow preventer
4. Pressure-regulating
5. Air-relief
6. Pressure-relief
7. Antisiphon
8. Quick-coupler

vegetative propagation—Asexual propagation using pieces of vegetation, i.e., sprigs or sod pieces

velocity—Speed of water in a pipe expressed in feet per second; not to be confused with pressure.

verdure—Layer of green living plant tissue remaining above the soil following mowing

vertical cutter—Powered mechanical device having vertically rotating blades or tines that cut into the face of a turf below the cutting height to control thatch and/or grain. The tine type is also referred to as a power rake.

vertical mower—Powered mower that cuts turf by high-speed impact of blades moving in a vertical plane; the blades can be of varied shapes and fixed or free-swinging (flail).

warm-season turfgrass—Turfgrass species best adapted to growth during the warmer part of the year; usually dormant during cold weather or injured by it; commonly having temperature optimums of 27° to 35°C, e.g., bahiagrass, bermudagrass, St. Augustinegrass, and zoysiagrass.

washboard effect—See **rippling**.

water demand—Same as water load

water hammer—Shock waves set up in pipelines due to surge condition

watering cycle—Refer to **operating cycle**.

watering requirements—Requirements demanded by the turf, including infiltration rate and infiltration capacity

watering-in—Watering turf immediately after application of chemicals to dissolve and/or wash materials from the plant surface into the soil

water load—Maximum capacity a system will require for operation expressed in gallons per minute

water region—Water resources region as designated by the U.S. Water Resources Council. There are 21 regions, 18 in the conterminous United States and one each for Alaska, Hawaii, and the Caribbean.

water subregion—Subdivision of a region. There are 106 subregions used exclusively in the Second National Water Assessment.

water table—Top level of permanent groundwater zone

wear—Collective direct injurious effects of traffic on a turf, as distinct from the indirect effects of traffic caused by soil compaction

wet wilt—Wilting of turf in the presence of free soil water when evapotranspiration exceeds water uptake by roots

whipping pole—Bamboo stalk or similar pole used in poling turf

windburn, turf—Death and browning, most commonly occurring on the uppermost leaves of grasses, caused by atmospheric desiccation

winter desiccation—Death of leaves or plants by drying during winter dormancy

winter discoloration—See **low temperature discoloration**.

winter fertilization—A late-fall-to-winter application of fertilizer to turfgrasses at rates which maintain green color without adverse physiological effects; used in regions characterized by moderate winters for the species involved.

winter overseeding—Seeding cool-season turfgrasses over warm-season turfgrasses at or near the start of winter dormancy; used in mild climates to provide green, growing turf during winter when warm-season species are brown and dormant.

winter-protection cover—Barrier placed over a turf to prevent winter desiccation, to insulate against low temperature stress, and to stimulate early spring greenup

winterkill—Any injury to turfgrass plants that occurs during winter

withdrawal—Water taken from a surface or groundwater source for off-stream use

working pressure—Pressure at any point of the system while the system is in operation